EP Math 5/6 Workbook

This book belongs to

Copyright © 2018 PuzzleFast Books. All rights reserved.

This workbook, made by permission of Easy Peasy All-in-One Homeschool, is based on the math component of Easy Peasy's curriculum. For EP's online curriculum visit allinonehomeschool.com.

This book may not be reproduced in whole or in part in any manner whatsoever without written permission from the publisher. For more information visit www.puzzlefast.com.

ISBN: 9798448621888

How to Use This Workbook

This is an offline workbook for Easy Peasy All-in-One Homeschool's Math 5/6 course. We've modified and expanded upon the online activities available at the Easy Peasy All-in-One Homeschool website so that your child can work offline if desired. You can use this book as a stand-alone course or use it along with the online course whenever extra practice would be useful or when you need to be away from the computer.

This workbook follows the EP math course in sequential order, providing 139 daily lessons with practice worksheets that can replace online lessons and exercises. The lessons are designed with the following objectives in mind:

- **To introduce an appropriate amount of new material each day**

    Each lesson is built upon the previous lessons and introduces new topics in a way that flows logically and makes sense. Your child will learn new skills and strategies each day without getting overwhelmed.

- **To teach by examples and by practice**

    Each lesson has two pages. The first page starts with the day's lesson presented in the form of examples, followed by practice problems for reinforcing the concepts taught. The second page provides additional practice problems.

    For each lesson, have your child work through the examples first. If your child has difficulty following them, provide help using the strategies described in your Parent's Guide. Having your child pretend to be a teacher and explain step-by-step how to solve the examples is another great way to build understanding of the topic. Use the second page as needed to make sure your child understands the topic.

- **To provide offline alternatives to online activities**

    Each lesson can replace the day's online activities. It provides offline activities comparable in both quantity and quality to those offered online, so that your child can switch between online and offline without missing anything.

As you proceed through the EP curriculum, use this workbook to exercise your child's math skills.

Where to Find the Answer Key

Please note:

- The solutions are on the EP website as well as in the Parent's Guide and are **not included** in this workbook.

Completion Chart

		LESSON	PRACTICE
LESSON 1	Basic Addition	☐	☐
LESSON 2	Basic Subtraction	☐	☐
LESSON 3	Adding 2-Digit Numbers with Regrouping	☐	☐
LESSON 4	Subtracting 2-Digit Numbers with Regrouping	☐	☐
LESSON 5	Adding 3-Digit Numbers with Regrouping	☐	☐
LESSON 6	Subtracting 3-Digit Numbers with Regrouping	☐	☐
LESSON 7	Alternative Method: Subtracting Left-to-Right	☐	☐
LESSON 8	Adding Multi-Digit Numbers	☐	☐
LESSON 9	Subtracting Across Zeros	☐	☐
LESSON 10	Word Problems: Adding/Subtracting Whole Numbers	☐	☐
LESSON 11	Introduction to Integers	☐	☐
LESSON 12	Ordering Integers	☐	☐
LESSON 13	Adding Integers	☐	☐
LESSON 14	Subtracting Integers	☐	☐
LESSON 15	Word Problems: Integers	☐	☐
LESSON 16	Adding and Subtracting Integers	☐	☐
LESSON 17	Basic Multiplication	☐	☐
LESSON 18	Multiplying by Multiples of 10	☐	☐
LESSON 19	Basic Division	☐	☐
LESSON 20	Long Division	☐	☐
LESSON 21	Multiplying 1- and 2-Digit Numbers	☐	☐
LESSON 22	Multiplying Multi-Digit Numbers	☐	☐
LESSON 23	Alternative Method: Lattice Multiplication	☐	☐
LESSON 24	Long Division with Remainders	☐	☐
LESSON 25	Word Problems: Multiplying/Dividing Whole Numbers	☐	☐
LESSON 26	Dividing by Multi-Digit Numbers	☐	☐
LESSON 27	Alternative Method: Partial Quotient Division	☐	☐
LESSON 28	Word Problems: Arithmetic with Whole Numbers	☐	☐
LESSON 29	Place Value	☐	☐

Easy Peasy All-in-One Homeschool   EP Math 5/6 Workbook

Lesson 30	Rounding Whole Numbers	☐	☐
Lesson 31	Multiplying and Dividing Integers	☐	☐
Lesson 32	Understanding Exponents	☐	☐
Lesson 33	Commutative Law of Addition and Multiplication	☐	☐
Lesson 34	Associative Law of Addition and Multiplication	☐	☐
Lesson 35	Distributive Law of Multiplication	☐	☐
Lesson 36	Absolute Value	☐	☐
Lesson 37	Comparing Absolute Values	☐	☐
Lesson 38	Identity Property of 0 and 1	☐	☐
Lesson 39	Inverse Property of Addition and Multiplication	☐	☐
Lesson 40	Dividing by Zero	☐	☐
Lesson 41	Arithmetic with Integers	☐	☐
Lesson 42	Order of Operations	☐	☒
Lesson 43	Prime Numbers	☐	☐
Lesson 44	Recognizing Prime Numbers	☐	☐
Lesson 45	Divisibility Rules	☐	☐
Lesson 46	Finding Factor Pairs	☐	☐
Lesson 47	Finding Factors	☐	☐
Lesson 48	Prime Factorization	☐	☐
Lesson 49	Least Common Multiple (LCM)	☐	☐
Lesson 50	Greatest Common Divisor (GCD)	☐	☐
Lesson 51	Word Problems: LCMs and GCDs	☐	☒
Lesson 52	Order of Operations with Parentheses	☐	☐
Lesson 53	Identifying Numerators and Denominators	☐	☐
Lesson 54	Identifying Fraction Parts	☐	☐
Lesson 55	Equivalent Fractions	☐	☐
Lesson 56	Word Problems: Equivalent Fractions	☐	☐
Lesson 57	Comparing Fractions	☐	☐
Lesson 58	Comparing Fractions	☐	☐
Lesson 59	Simplifying Fractions	☐	☐
Lesson 60	Ordering Fractions	☐	☐
Lesson 61	Types of Fractions	☐	☐

Easy Peasy All-in-One Homeschool   EP Math 5/6 Workbook

Lesson 62	Mixed Numbers to Improper Fractions	☐	☐
Lesson 63	Improper Fractions to Mixed Numbers	☐	☐
Lesson 64	Comparing Mixed Numbers	☐	☐
Lesson 65	Comparing Mixed and Improper Fractions	☐	☐
Lesson 66	Adding Fractions with Like Denominators	☐	☐
Lesson 67	Subtracting Fractions with Like Denominators	☐	☐
Lesson 68	Finding the Least Common Multiple	☐	☐
Lesson 69	Finding Common Denominators	☐	☐
Lesson 70	Adding Fractions with Unlike Denominators	☐	☐
Lesson 71	Subtracting Fractions with Unlike Denominators	☐	☐
Lesson 72	Word Problems: Adding/Subtracting Fractions	☐	☐
Lesson 73	Adding Negative Fractions	☐	☐
Lesson 74	Subtracting Negative Fractions	☐	☐
Lesson 75	Fractions on a Number Line	☐	☐
Lesson 76	Fractions on a Number Line	☐	☐
Lesson 77	Negative Fractions on a Number Line	☐	☐
Lesson 78	Adding Mixed Numbers	☐	☐
Lesson 79	Subtracting Mixed Numbers	☐	☐
Lesson 80	Word Problems: Adding/Subtracting Mixed Numbers	☐	☐
Lesson 81	Multiplying Fractions	☐	☐
Lesson 82	Multiplying Mixed Numbers	☐	☐
Lesson 83	Multiplying Fractions of All Types	☐	☐
Lesson 84	Word Problems: Multiplying Fractions	☐	☐
Lesson 85	Dividing Fractions	☐	☐
Lesson 86	Word Problems: Dividing Fractions	☐	☐
Lesson 87	Reciprocal of a Mixed Number	☐	☐
Lesson 88	Dividing Fractions of All Types	☐	☐
Lesson 89	Decimal Place Value	☐	☐
Lesson 90	Decimals on a Number Line	☐	☐
Lesson 91	Rounding Decimals	☐	☐
Lesson 92	Estimation with Decimals	☐	☐
Lesson 93	Comparing Decimals	☐	☐

Lesson 94	Adding Decimals	☐	☐
Lesson 95	Adding Decimals	☐	☐
Lesson 96	Adding Decimals	☐	☐
Lesson 97	Subtracting Decimals	☐	☐
Lesson 98	Subtracting Decimals	☐	☐
Lesson 99	Word Problems: Adding/Subtracting Decimals	☐	☐
Lesson 100	Multiplying Decimals by Powers of 10	☐	☐
Lesson 101	Multiplying Decimals	☐	☐
Lesson 102	Dividing Decimals by Powers of 10	☐	☐
Lesson 103	Dividing Decimals	☐	☐
Lesson 104	Dividing Decimals	☐	☐
Lesson 105	Converting Fractions to Decimals	☐	☐
Lesson 106	Converting Fractions to Decimals	☐	☐
Lesson 107	Converting Decimals to Fractions	☐	☐
Lesson 108	Converting Decimals to Fractions	☐	☐
Lesson 109	Converting Fractions to Repeating Decimals	☐	☐
Lesson 110	The Meaning of Percent	☐	☐
Lesson 111	Converting Decimals to Percents	☐	☐
Lesson 112	Converting Percents to Decimals	☐	☐
Lesson 113	Fractions and Decimals on a Number Line	☐	☐
Lesson 114	Ordering Numeric Expressions	☐	☐
Lesson 115	Numbers as Fractions, Decimals, and Percents	☐	☐
Lesson 116	Identifying Percent, Whole, and Part	☐	☐
Lesson 117	Types of Percent Problems	☐	☐
Lesson 118	Word Problems: Percents	☐	☐
Lesson 119	Introduction to Ratios	☐	☐
Lesson 120	Simplifying Ratios	☐	☐
Lesson 121	Simplifying Ratios	☐	☐
Lesson 122	Word Problems: Equivalent Ratios	☐	☐
Lesson 123	Understanding Proportions	☐	☐
Lesson 124	Solving Proportions	☐	☐
Lesson 125	Solving Proportions	☐	☐

Easy Peasy All-in-One Homeschool   EP Math 5/6 Workbook

LESSON 126	Finding Unit Rates	☐	☐
LESSON 127	Finding Unit Prices	☐	☐
LESSON 128	Converting Units	☐	☐
LESSON 129	Converting Units of Length	☐	☐
LESSON 130	Customary Units and Metric Units	☐	☐
LESSON 131	Converting between Metric Units	☐	☐
LESSON 132	Converting Customary Units of Volume	☐	☐
LESSON 133	Converting Customary Units of Weight	☐	☐
LESSON 134	Converting Temperatures	☐	☐
LESSON 135	Converting Customary and Metric Units	☐	☐
LESSON 136	Exponents with Integer Bases	☐	☐
LESSON 137	Patterns in Zeros	☐	☐
LESSON 138	Order of Operations with Exponents	☐	☐
LESSON 139	Understanding Square Roots	☐	☐

LESSON 1 Basic Addition

A. You can count to add. Put four coins on the table. Now put three more coins. How many coins are on the table now?

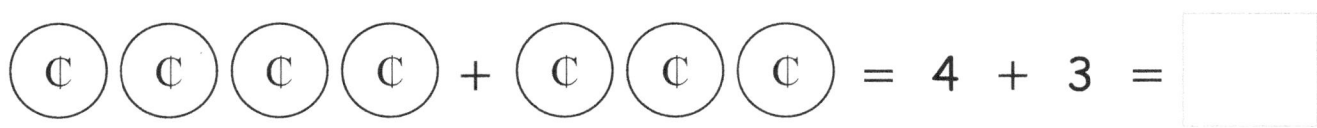 = 4 + 3 =

B. You can also use a number line to add. Put your finger on 3. Jump four numbers to the right. That adds four. What number are you on now?

 3 + 4 =

C. Add numbers up to 9.

8 + 6 =	7 + 7 =	6 + 9 =
5 + 7 =	9 + 1 =	5 + 4 =
8 + 9 =	6 + 2 =	4 + 3 =
6 + 6 =	4 + 9 =	8 + 2 =
2 + 5 =	7 + 3 =	3 + 6 =
7 + 6 =	6 + 4 =	9 + 9 =
2 + 9 =	8 + 8 =	7 + 4 =
5 + 5 =	3 + 5 =	4 + 2 =

D. Add 2-digit numbers by adding up to 9 in each place value.

```
   32         23         24         25         37
+ 43       + 22       + 64       + 34       + 60
-----      -----      -----      -----      -----
   75
```

Easy Peasy All-in-One Homeschool   EP Math 5/6 Workbook

LESSON 1 Practice

A. Add numbers up to 9.

$3 + 1 =$ $4 + 1 =$ $0 + 0 =$

$2 + 3 =$ $1 + 2 =$ $2 + 2 =$

$7 + 9 =$ $6 + 3 =$ $6 + 8 =$

$5 + 6 =$ $9 + 9 =$ $7 + 5 =$

$9 + 6 =$ $4 + 7 =$ $9 + 8 =$

$3 + 7 =$ $2 + 4 =$ $6 + 6 =$

$4 + 6 =$ $9 + 3 =$ $4 + 8 =$

$8 + 8 =$ $6 + 7 =$ $5 + 9 =$

$5 + 3 =$ $9 + 2 =$ $7 + 7 =$

$4 + 4 =$ $5 + 5 =$ $1 + 9 =$

$4 + 5 =$ $8 + 3 =$ $3 + 4 =$

$8 + 7 =$ $0 + 5 =$ $2 + 6 =$

$2 + 8 =$ $2 + 7 =$ $9 + 4 =$

$3 + 3 =$ $5 + 2 =$ $8 + 5 =$

B. Add 2-digit numbers without regrouping (carrying).

$$\begin{array}{r} 22 \\ +\ 43 \\ \hline \end{array} \qquad \begin{array}{r} 23 \\ +\ 53 \\ \hline \end{array} \qquad \begin{array}{r} 56 \\ +\ 31 \\ \hline \end{array} \qquad \begin{array}{r} 35 \\ +\ 32 \\ \hline \end{array} \qquad \begin{array}{r} 63 \\ +\ 26 \\ \hline \end{array}$$

Easy Peasy All-in-One Homeschool   EP Math 5/6 Workbook

LESSON 2 Basic Subtraction

A. You can count to subtract. Put seven coins on the table. Now take away four of the coins. How many coins are on the table now?

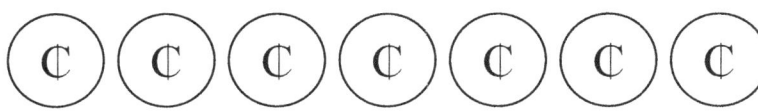 7 − 4 =

B. You can also use a number line to subtract. Put your finger on 7. Jump three numbers to the left. That subtracts three. What number are you on now?

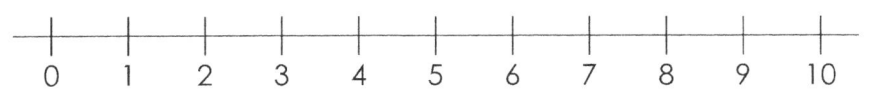 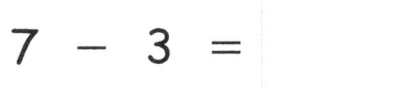 7 − 3 =

C. Subtract numbers up to 9.

15 − 9 =	10 − 1 =	14 − 6 =
9 − 4 =	14 − 7 =	12 − 7 =
15 − 8 =	8 − 2 =	17 − 9 =
10 − 2 =	13 − 9 =	12 − 6 =
9 − 6 =	10 − 4 =	11 − 8 =
18 − 9 =	14 − 5 =	13 − 6 =
11 − 4 =	16 − 8 =	7 − 3 =
12 − 5 =	17 − 8 =	15 − 7 =

D. Subtract 2-digit numbers by subtracting each place value.

```
   79        59        86        97        68
 − 52      − 43      − 54      − 53      − 32
 ────
   27
```

Easy Peasy All-in-One Homeschool   EP Math 5/6 Workbook

LESSON 2 Practice

A. Subtract numbers up to 9.

5 − 1 =	4 − 2 =	5 − 5 =
4 − 4 =	3 − 1 =	5 − 3 =
14 − 8 =	9 − 3 =	16 − 9 =
12 − 5 =	18 − 9 =	11 − 6 =
8 − 4 =	11 − 7 =	15 − 6 =
14 − 7 =	6 − 4 =	10 − 7 =
12 − 8 =	10 − 6 =	12 − 3 =
14 − 9 =	13 − 7 =	16 − 8 =
12 − 6 =	11 − 2 =	8 − 3 =
13 − 4 =	10 − 5 =	17 − 9 =
10 − 9 =	9 − 7 =	11 − 3 =
8 − 6 =	5 − 5 =	9 − 5 =
15 − 9 =	17 − 8 =	15 − 7 =
13 − 5 =	12 − 9 =	10 − 8 =

B. Subtract 2-digit numbers without regrouping (borrowing).

```
   95        58        67        79        86
 − 42      − 45      − 34      − 22      − 34
 ────      ────      ────      ────      ────
```

Easy Peasy All-in-One Homeschool EP Math 5/6 Workbook

LESSON 3 Adding 2-Digit Numbers with Regrouping

A. Here are the steps for adding 2-digit numbers with regrouping (carrying).

```
  1
  2 8
+ 4 5
-----
  7 3
```

1. Add the ones digits. 8 + 5 = 13
2. Write 3 in the ones column and carry the one.
3. Add the tens digits. 1 (carryover) + 2 + 4 = 7
4. Write 7 in the tens column.

B. You can also add each place value separately, and then add the sums.

```
  20 + 8
+ 40 + 5
---------
  60 + 13 = 73
```

1. Think about both numbers in expanded form.
2. Add the tens. 20 + 70 = 90
3. Add the ones. 3 + 5 = 8
4. Add the tens and ones sums. 90 + 8 = 98

C. Add using place value. You can see an example below of using place value.

```
   15        17        36         9        28
 + 26      + 46      + 18      + 35      + 54
 ----
   30
 + 11
 ----
   41
```

D. Add using regrouping (carrying).

```
   1
   38        23        27        45        37
 + 43      + 28      + 64      + 39      + 58
 ----
   81
```

E. Add using the method of your choice.

```
   22        26        56        35        49
 + 49      + 54      + 37      + 38      + 26
```

Easy Peasy All-in-One Homeschool   EP Math 5/6 Workbook

LESSON 3 Practice

A. Add numbers up to 20.

$5 + 14 =$ $13 + 6 =$ $8 + 9 =$

$13 + 3 =$ $5 + 7 =$ $3 + 14 =$

$9 + 6 =$ $14 + 4 =$ $5 + 5 =$

$6 + 12 =$ $8 + 7 =$ $17 + 2 =$

$7 + 6 =$ $3 + 15 =$ $12 + 5 =$

B. Add the 2-digit numbers.

```
  44        25        53        28        26        47
+ 38      + 62      + 37      + 58      + 75      + 14
----      ----      ----      ----      ----      ----


  32        45        79        34        27        69
+ 25      + 25      + 25      + 18      + 26      + 39
----      ----      ----      ----      ----      ----


  40        56        58        69        24        46
+ 20      + 14      + 37      + 25      + 58      + 45
----      ----      ----      ----      ----      ----
```

Easy Peasy All-in-One Homeschool   EP Math 5/6 Workbook

LESSON 4 Subtracting 2-Digit Numbers with Regrouping

A. Here are the steps for subtracting 2-digit numbers with regrouping (borrowing).

```
  8 13
  9̸ 3̸
+ 2 8
-----
  6 5
```

1. Subtract the ones column. 3 − 8 requires regrouping.
2. Borrow 1 from the tens place. 9 becomes 8. 3 becomes 13.
3. Subtract the ones column again. 13 − 8 = 5
4. Write 5 in the ones column.
5. Subtract the tens column. 8 − 2 = 6
6. Write 6 in the tens column.

B. Subtract the 2-digit numbers.

```
  7 10
   8̸0̸         52          75          94          87
−  35       − 18        − 67        − 58        − 23
-----       ----        ----        ----        ----
   45
```

```
   91          84          82          68          70
−  25       − 46        − 19        − 25        − 31
-----       ----        ----        ----        ----
```

```
   74          57          61          60          85
−  65       − 14        − 38        − 45        − 29
-----       ----        ----        ----        ----
```

A math riddle for you! If 3 cats catch 3 mice in 3 minutes, how long will it take 10 cats to catch 10 mice?

Easy Peasy All-in-One Homeschool   EP Math 5/6 Workbook

LESSON 4 Practice

A. Subtract numbers up to 20.

16 − 8 =   12 − 6 =   15 − 6 =

13 − 4 =   11 − 7 =   11 − 6 =

14 − 8 =   17 − 9 =   18 − 9 =

10 − 2 =   14 − 5 =   10 − 5 =

11 − 5 =   13 − 8 =   12 − 9 =

B. Subtract the 2-digit numbers.

```
  62        57        83        78        94        56
− 29      − 14      − 45      − 59      − 37      − 21



  55        60        42        76        85        93
− 46      − 38      − 16      − 49      − 52      − 27



  80        46        79        68        95        84
− 65      − 28      − 24      − 22      − 78      − 39
```

LESSON 5 Adding 3-Digit Numbers with Regrouping

A. Here are the steps for adding 3-digit numbers with regrouping (carrying).

```
  1 1
  7 8 3
+ 5 6 9
-------
1 3 5 2
```

1. Add the ones digits. 3 + 9 = 12
2. Write 2 in the ones column and carry 1 to the tens column.
3. Add the tens digits. 1 (carryover) + 8 + 6 = 15
4. Write 5 in the tens column and carry 1 to the hundreds column.
5. Add the hundreds digits. 1 (carryover) + 7 + 5 = 13
6. Write 1 in the thousands column and 3 in the hundreds column.

B. You can also add each place value separately, and then add the sums.

$$\begin{array}{r} 700 + 80 + 3 \\ + 500 + 60 + 9 \\ \hline 1200 + 140 + 12 \\ = 1352 \end{array}$$

1. Think about both numbers in expanded form.
2. Add the hundreds. 700 + 500 = 1200
3. Add the tens. 80 + 60 = 140
4. Add the ones. 3 + 9 = 12
5. Add the hundreds, tens, and ones sums.
   1200 + 140 + 12 = 1352

C. Add the 3-digit numbers using place value. See the example.

```
   357          948          873          214
+  258        + 675        + 865        + 368
------        -----        -----        -----
   500
   100
+   15
------
   615
```

D. Add the 3-digit numbers using regrouping (carrying).

```
   274          924          965          479
+  554        + 547        + 364        + 677
------        -----        -----        -----
```

Easy Peasy All-in-One Homeschool   EP Math 5/6 Workbook

LESSON 5 Practice

A. Add the 2-digit numbers.

```
   83        81        65        27        38
 + 19      + 69      + 23      + 43      + 58
 ----      ----      ----      ----      ----
```

```
   45        42        28        59        43
 + 89      + 67      + 67      + 49      + 26
 ----      ----      ----      ----      ----
```

B. Add the 3-digit numbers.

```
   547       484       573       329
 + 295     + 176     + 459     + 416
 -----     -----     -----     -----
```

```
   487       524       250       764
 + 354     + 476     + 594     + 942
 -----     -----     -----     -----
```

```
   548       746       148       403
 + 258     + 557     + 390     + 296
 -----     -----     -----     -----
```

Easy Peasy All-in-One Homeschool　　　　　　　　　　EP Math 5/6 Workbook

LESSON 6 Subtracting 3-Digit Numbers with Regrouping

A. Here are the steps for subtracting 3-digit numbers with regrouping (borrowing).

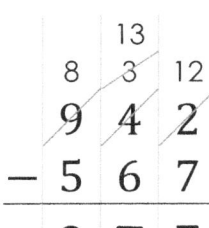

1. Subtract the ones column. 2 − 7 requires regrouping.
2. Borrow 1 from the tens place. 4 becomes 3. 2 becomes 12.
3. Subtract the ones column again. 12 − 7 = 5
4. Write 5 in the ones column.
5. Subtract the tens column. 3 − 6 requires regrouping.
6. Borrow 1 from the hundreds place. 9 becomes 8. 3 becomes 13.
7. Subtract the tens column again. 13 − 6 = 7
8. Write 7 in the tens column.
9. Subtract the hundreds column. 8 − 5 = 3
10. Write 3 in the hundreds column.

B. Subtract the 3-digit numbers. You can work out the problems in your notebook. A tip is to turn the notebook sideways and use the lines on the paper to keep the digits aligned.

923	782	952	834
− 832	− 206	− 287	− 562

679	831	590	628
− 324	− 257	− 453	− 565

826	735	942	735
− 467	− 597	− 689	− 358

Did you know? The first number to contain the letter 'a' is one thousand.

LESSON 6 Practice

A. Subtract the 2-digit numbers.

```
  74        72        75        63        29
- 58      - 27      - 45      - 49      - 25
```

```
  84        96        60        95        67
- 29      - 56      - 18      - 63      - 30
```

```
  75        83        34        63        92
- 37      - 50      - 19      - 25      - 38
```

B. Subtract the 3-digit numbers.

```
  758       910       679       836
- 392     - 447     - 593     - 483
```

```
  527       764       842       534
- 235     - 348     - 274     - 419
```

LESSON 7 Alternative Method: Subtracting Left-to-Right

A. Follow these examples to understand why borrowing works.

$$842 = 800 + 40 + 2 = 800 + 40 + 2$$
$$-\ 64 = -\ 60 - 4 = -\ 60 - 4$$

With borrowing: 40 → 30, 2 → 12.

You can't subtract 4 from 2.
You borrow 10 from 40.
40 becomes 40 − 10 = 30.
2 becomes 2 + 10 = 12.
You can subtract 4 from 12.

With further borrowing: 800 → 700, 30 → 130.

You can't subtract 60 from 30.
You borrow 100 from 800.
800 becomes 800 − 100 = 700.
30 becomes 30 + 100 = 130.
You can subtract 60 from 130.

B. There are many ways to subtract numbers. The following example illustrates an alternative subtraction method, subtracting from left to right. Complete the example.

```
    1
  9 4 7 2
− 5 6 3 9
─────────
  3 8
```

1. Subtract the thousands. 9 − 5 = 4
2. Look at the hundreds column. It needs borrowing. Give 1 to the hundreds column and write 4 − 1 = 3 in the answer.
3. Subtract the hundreds with the borrowed 10. 14 − 6 = 8
4. The tens column doesn't need borrowing, so just write 8.
5. Continue subtracting and borrowing to the next column.

C. Subtract the 4-digit numbers. Use whichever method you like.

6425	8573	5930	7238
− 3798	− 4935	− 2665	− 3693

9215	7355	8290	7620
− 4487	− 4636	− 2721	− 6278

Easy Peasy All-in-One Homeschool   EP Math 5/6 Workbook

LESSON 7 Practice

A. Subtract numbers up to 100.

72	97	82	90	75	68
− 39	− 88	− 74	− 59	− 27	− 26

93	84	68	96	62	73
− 55	− 57	− 24	− 68	− 46	− 59

97	65	49	54	83	84
− 49	− 28	− 17	− 29	− 37	− 26

B. Subtract the 4-digit numbers.

8752	9455	7422	8150	6250
− 3434	− 2827	− 4849	− 2587	− 3594

4134	8577	5930	7273	8356
− 1689	− 4939	− 1675	− 3698	− 5762

Lesson 8 Adding Multi-Digit Numbers

A. You can add any large numbers using regrouping. Complete the example.

```
     1 1
  7 3 5 4 9 8
+ 7 9 8 6 8 6
  _____
          8 4
```

1. Add the ones digits. 8 + 6 = 14
2. Write 4 in the ones column and carry 1 to the tens column.
3. Add the tens digits. 1 (carryover) + 9 + 8 = 18
4. Write 8 in the tens column and carry 1 to the hundreds column.
5. Continue adding and carrying to the next column.

B. Add the multi-digit numbers.

```
   9785          6248          6249          4869
+  3208       +  6578       +  3928       +  7487
```

```
   8347          6265          8465          4392
+  3607       +  4953       +  7453       +  2358
```

```
   538,532       1,365,238     4,527,648
+  207,886    +    257,494  +    936,561
```

A math riddle for you! How can you add eight 8's to get the number 1000?

Easy Peasy All-in-One Homeschool   EP Math 5/6 Workbook

LESSON 8 Practice

A. Add numbers up to 1000.

```
   753        580        839        237        944
+  783     +  847     +  179     +  946     +  986
```

```
   547        769        542        243        247
+  628     +  965     +  958     +  632     +  794
```

```
   462        528        489        539        958
+  984     +  785     +  469     +  456     +  373
```

B. Add the multi-digit numbers.

```
   3205       9278       8968       7868       4352
+  8546    +  6458    +  7949    +  4762    +  5787
```

```
   863,629      437,558     2,468,593     4,699,073
+  562,760   +  857,458   +   958,495   +   259,546
```

Easy Peasy All-in-One Homeschool   EP Math 5/6 Workbook

LESSON 9 Subtracting Across Zeros

A. Here are the steps for subtracting 3-digit numbers across zeros. Complete the example.

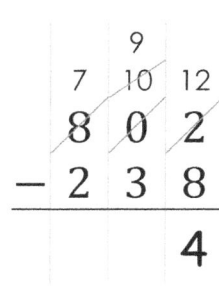

1. Subtract the ones column. 2 − 8 requires regrouping.
2. Borrow 1 from the tens place. But there's nothing to borrow.
3. Borrow 1 from the hundreds place. 8 becomes 7. 0 becomes 10.
4. Now borrow 1 from the tens place. 10 becomes 9. 2 becomes 12.
5. Subtract the ones column. 12 − 2 = 4
6. Write 4 in the ones column.
7. Continue subtracting to the next column.

B. You can also use the box method, as shown below. Complete the example.

```
 5  9  9  13
⎡6  0  0⎤ 3
⎣       ⎦
−  4  7  5   8
─────────────
             5
```

1. Subtract the ones column. 3 − 8 requires regrouping.
2. There are no tens. Find the first non-zero digit in the columns to the left of the tens place. Draw a box from that digit to the tens place. The first non-zero digit is 6, so the box goes around 600.
3. Take 1 from 600. 600 becomes 599. 3 becomes 13.
4. Subtract as usual from right to left, column by column.

C. Subtract the numbers across zeros.

```
   505          700          802          900
 − 238        − 482        − 357        − 284
 —————        —————        —————        —————


   900          604          800          502
 − 529        − 279        − 463        − 236
 —————        —————        —————        —————


  7000         3000         9003         4000
− 1650       − 2350       − 5080       − 2710
 —————        —————        —————        —————
```

Easy Peasy All-in-One Homeschool   EP Math 5/6 Workbook

LESSON 9 Practice

A. Subtract numbers up to 1000.

$$\begin{array}{r} 854 \\ -299 \\ \hline \end{array} \qquad \begin{array}{r} 713 \\ -497 \\ \hline \end{array} \qquad \begin{array}{r} 562 \\ -229 \\ \hline \end{array} \qquad \begin{array}{r} 865 \\ -583 \\ \hline \end{array} \qquad \begin{array}{r} 931 \\ -368 \\ \hline \end{array}$$

$$\begin{array}{r} 615 \\ -338 \\ \hline \end{array} \qquad \begin{array}{r} 786 \\ -547 \\ \hline \end{array} \qquad \begin{array}{r} 980 \\ -286 \\ \hline \end{array} \qquad \begin{array}{r} 747 \\ -289 \\ \hline \end{array} \qquad \begin{array}{r} 812 \\ -367 \\ \hline \end{array}$$

$$\begin{array}{r} 934 \\ -568 \\ \hline \end{array} \qquad \begin{array}{r} 690 \\ -259 \\ \hline \end{array} \qquad \begin{array}{r} 834 \\ -297 \\ \hline \end{array} \qquad \begin{array}{r} 753 \\ -376 \\ \hline \end{array} \qquad \begin{array}{r} 956 \\ -424 \\ \hline \end{array}$$

B. Subtract the numbers across zeros.

$$\begin{array}{r} 803 \\ -298 \\ \hline \end{array} \qquad \begin{array}{r} 504 \\ -276 \\ \hline \end{array} \qquad \begin{array}{r} 900 \\ -454 \\ \hline \end{array} \qquad \begin{array}{r} 605 \\ -257 \\ \hline \end{array} \qquad \begin{array}{r} 700 \\ -319 \\ \hline \end{array}$$

$$\begin{array}{r} 900 \\ -393 \\ \hline \end{array} \qquad \begin{array}{r} 403 \\ -227 \\ \hline \end{array} \qquad \begin{array}{r} 505 \\ -238 \\ \hline \end{array} \qquad \begin{array}{r} 802 \\ -445 \\ \hline \end{array} \qquad \begin{array}{r} 704 \\ -229 \\ \hline \end{array}$$

Easy Peasy All-in-One Homeschool   EP Math 5/6 Workbook

LESSON 10 Word Problems: Adding/Subtracting Whole Numbers

A. There are multiple ways to solve word problems. The following examples show just a few of them. It is okay to use your own strategy.

Mia has 90 pennies, nickels, and dimes. Twenty-eight of them are dimes and 45 are nickels. How many pennies does Mia have? **17 pennies**

Solve the problem:
28 dimes + 45 nickels = 73
# of pennies = 90 − 73 = 17

Check your answer:
28 + 45 + 17 = 90

Ryan had some marbles. He lost 19 of them and found 7. If he had 62 after that, how many marbles did Ryan have at first? **74 marbles**

Solve the problem:
Before finding 7 marbles: 62 − 7 = 55
Before losing 19 marbles: 55 + 19 = 74

Check your answer:
74 − 19 + 7 = 62

B. Solve the word problems. Make sure to check your answers.

Sarah has 62 red, blue, and yellow markers. 15 of them are red and 19 are blue. How many markers are yellow? _____

Laura bought 55 paper cups. After using 48 cups, she bought 25 more cups. How many paper cups did she have in the end? _____

Ella planted 32 roses, 26 daisies, and 14 violets. Fifty-six of them have bloomed so far. How many flowers have not bloomed yet? _____

Ron has $49. Matt has $19 more than Ron. Naomi has $12 less than Matt. How much money do they have all together? _____

There were some ducks swimming in the pond. Fifteen of them flew away, and another 17 ducks flew in. If there were 24 ducks after that, how many ducks were swimming in the pond at first? _____

Mom bought some apples. She used 18 of them to make pies and 37 to make jam. If there were 14 apples left, how many apples did Mom buy at first? _____

Did you know? A paper cannot be folded more than 9 times. Try it yourself!

Easy Peasy All-in-One Homeschool EP Math 5/6 Workbook

LESSON 10 Practice

Solve the word problems. Make sure to check your answers.

Ron had 82 stickers. He gave 28 of them to Mark, 26 to Larry, and the rest to Kyle. How many stickers did Kyle get?

Sam wants to buy three books that cost $13, $12, and $18. He has $25. How much more money does Sam need?

The store had 71 bags of potatoes. It sold 55 bags and brought in 36 more bags. How many bags did the store have in the end?

Angela had some math problems to solve. She solved 28 problems yesterday and 29 today. She still has 11 problems left. How many problems did she have at first?

The pet store had 90 fish. After selling 67 fish, it brought in 55 more fish. How many fish did the store have in the end?

There were some people in the auditorium. Twenty-six of them left, and 15 people entered. If there were 48 people after that, how many people were in the auditorium at first?

The pet store had 85 goldfish and 70 angelfish. After selling 57 goldfish, it brought in 32 more angelfish. How many goldfish and angelfish did the store have in the end?

Claire found some seashells on the beach. She used 22 to decorate her room and gave 26 to her sister. If there were 18 seashells left, how many seashells did Clair find at first?

Kate rode her bike 11 miles to the library and then 14 miles to the park. Later she came home the same way. How many miles did Kate ride in total?

In the afternoon, Adam took an hour break from his studying. After spending 25 minutes listening to music and 20 minutes watching TV, he went for a walk. How many minutes did he walk during the break?

Easy Peasy All-in-One Homeschool

EP Math 5/6 Workbook

LESSON 11 Introduction to Integers

A. Fill in each blank with one of the two word choices: positive or negative.

Integers are numbers that are positive or negative. A _____ number is any number greater than zero. A _____ number is any number less than zero. Zero is neither positive nor negative. A _____ number is written with a negative sign (-) in front of it. If a number has no sign or a positive sign (+), it is a _____ number. On a number line, _____ numbers are located to the right of zero while _____ numbers are to the left of zero.

B. If you subtract a number from a smaller number, you get a negative answer. Put your finger on 3. Jump seven numbers to the left. What number are you on now?

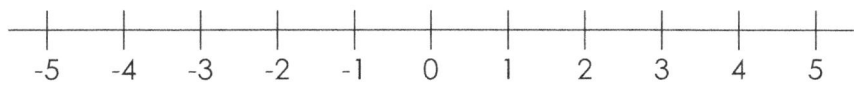

3 − 7 = _____

C. Use the number line above to subtract a number from a smaller number.

2 − 6 =   0 − 3 =   4 − 9 =

4 − 7 =   1 − 6 =   3 − 5 =

D. Fill in the missing numbers on each number line.

Easy Peasy All-in-One Homeschool   EP Math 5/6 Workbook

LESSON 11 Practice

A. Use the number line below to subtract a number from a smaller number.

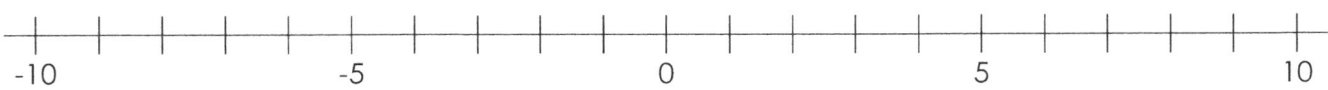

$4 - 9 =$ $6 - 10 =$ $8 - 15 =$

$0 - 6 =$ $9 - 17 =$ $5 - 14 =$

$1 - 8 =$ $3 - 12 =$ $7 - 13 =$

$2 - 5 =$ $5 - 11 =$ $9 - 18 =$

B. Fill in the missing numbers on each number line.

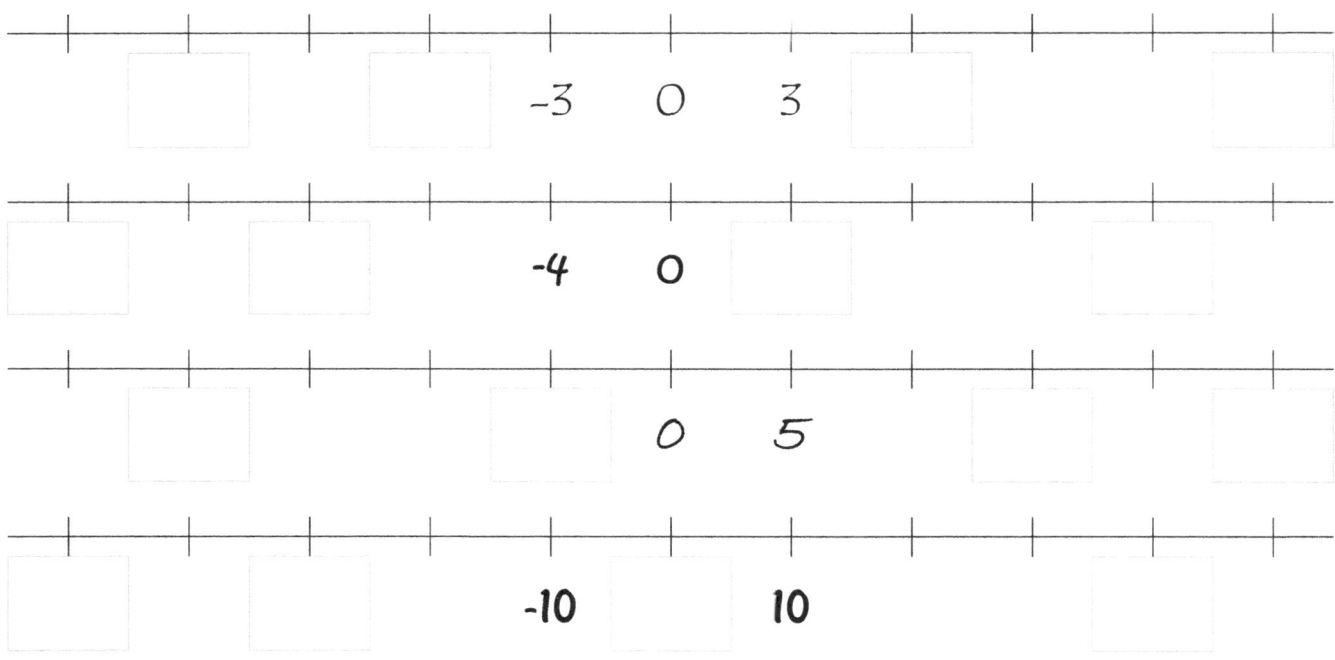

C. Use the number lines above to find the opposite values.

Opposite of $6 =$ Opposite of $-15 =$

Opposite of $-8 =$ Opposite of $40 =$

Did you know? Zero is the only number known by many different names, including cipher, null, naught, nil, zed, zilch, and zip.

Easy Peasy All-in-One Homeschool   EP Math 5/6 Workbook

LESSON 12 Ordering Integers

A. You can compare and order integers using a number line. When numbers are marked on a number line, smaller numbers are farther to the left. Here is an example of ordering numbers from least to greatest (from left to right on a number line).

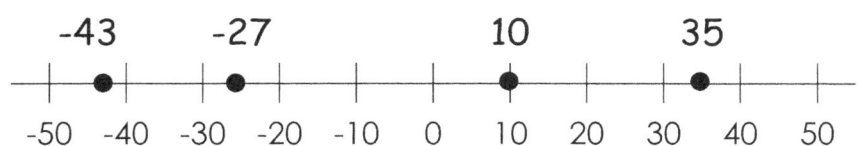

-27, 10, 35, -43

-43, -27, 10, 35

B. Place and label each set of numbers on the number line. Then write in the box the numbers from least to greatest.

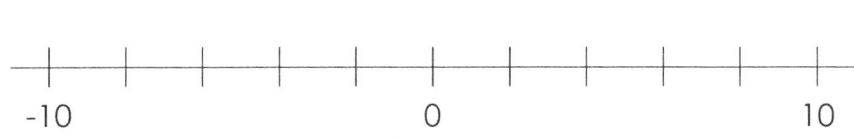

0, 2, -3, 6, -7

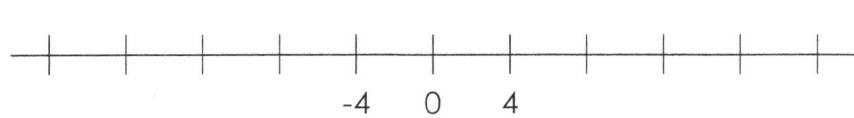

8, -2, -16, 14

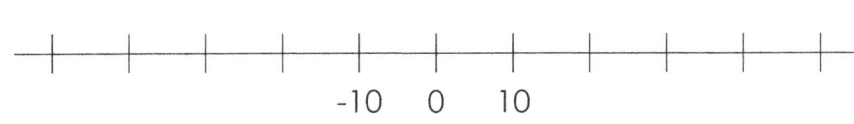

-28, 32, -50, -15

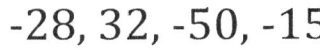

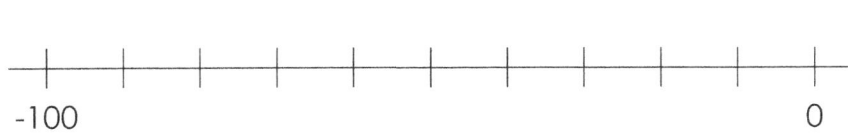

-61, -40, -27, -95

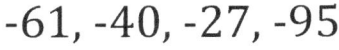

C. Order the numbers from greatest to least.

-5, 6, 2   0, -3, 8   -1, -9, -4, -7

_____   _____   _____

-10, -17, 13   -31, -62, -45   26, -38, 11, -57

_____   _____   _____

Easy Peasy All-in-One Homeschool   EP Math 5/6 Workbook

LESSON 12 Practice

A. Place and label each set of numbers on the number line. Then write in the box the numbers from least to greatest.

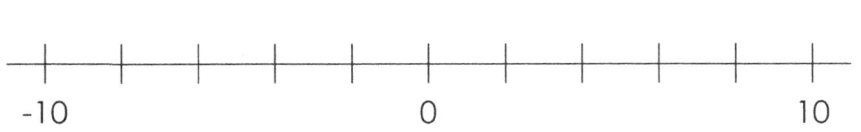

-1, -4, 3, -9, 8

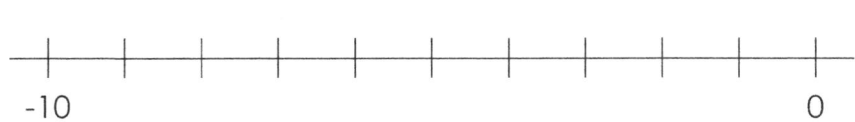

-7, -2, -6, -9

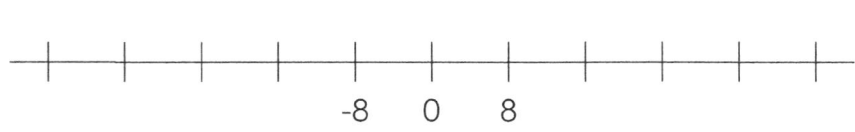

-12, 6, 38, -24

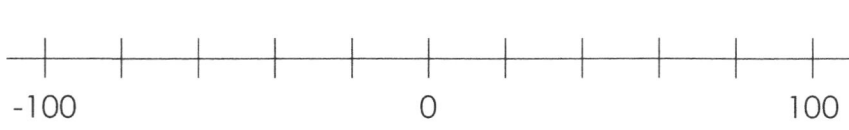

74, -80, 45, -16

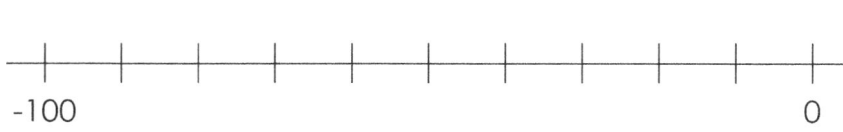

-71, -60, -94, -32

B. Order the numbers from greatest to least.

-7, 3, -4   2, -5, 0   -3, -5, -8, -1

_____   _____   _____

-10, 16, -18   -20, -75, -45 -86, -24, 36, -59

_____   _____   _____

A math riddle for you! I am an odd number. Take away one letter and I become even. What number am I?

Easy Peasy All-in-One Homeschool   EP Math 5/6 Workbook

LESSON 13 Adding Integers

A. To add numbers with different signs, move either to the right or to the left on a number line. Move to the right to add positive numbers. Move to the left to add negative numbers. Here are some examples. Use the number line below to follow along with each example.

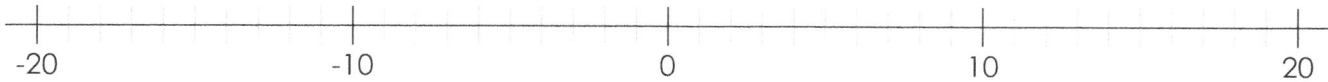

(-3) + (-9) = _____
Start at -3. Move 9 to the left to add -9. You land on -12. You added 9 and 3 in the negative direction.

5 + (-7) = _____
Start at 5. Move 7 to the left to add -7. You land on -2. You moved in opposite directions, in effect subtracting 7 from 5.

(-2) + 8 = _____
Start at -2. Move 8 to the right to add 8. You land on 6. Again moving in opposite directions, you in effect found 8 − 2.

2 + (-9) + 4 = _____
Start at 2. Move 9 to the left, and then 4 to the right. You land on -3. In effect you are adding 2 and 4 and subtracting 9.

B. Add the negative numbers. Use the number line above to help you.

(-2) + (-7) =   (-4) + (-7) =   (-8) + (-8) =

(-8) + (-9) =   (-7) + (-8) =   (-7) + (-3) =

(-4) + (-4) =   (-5) + (-2) =   (-8) + (-5) =

C. Add the numbers with different signs. Use the number line above to help you.

8 + (-6) =   4 + (-9) =   (-4) + 8 =

5 + (-1) =   2 + (-8) =   (-6) + 5 =

7 + (-4) =   5 + (-7) =   (-3) + 3 =

2 + (-6) + 5 =   (-4) + 9 + (-2) =

(-4) + 3 + 7 =   6 + (-4) + (-8) =

(-5) + (-8) + 6 =   (-4) + (-9) + (-5) =

Easy Peasy All-in-One Homeschool   EP Math 5/6 Workbook

LESSON 13 Practice

Add the numbers with different signs. Use the number line below to help you.

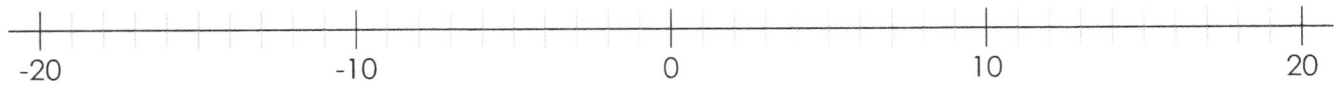

(-6) + (-3) =   (-5) + (-5) =   (-6) + (-9) =

(-4) + (-8) =   (-9) + (-7) =   (-7) + (-6) =

(-5) + (-6) =   (-3) + (-8) =   (-5) + (-3) =

6 + (-6) =   5 + (-9) =   (-2) + 7 =

8 + (-4) =   1 + (-6) =   (-8) + 5 =

9 + (-3) =   7 + (-8) =   (-5) + 9 =

(-7) + 9 =   8 + (-2) =   (-8) + 3 =

7 + (-4) =   (-6) + (-4) =   3 + (-6) =

(-4) + (-9) =   (-7) + 7 =   (-6) + (-8) =

6 + (-5) =   (-7) + (-7) =   (-9) + 2 =

(-8) + 4 + 5 =   6 + 3 + (-9) =

2 + (-7) + 9 =   (-4) + 6 + (-8) =

(-5) + (-6) + 7 =   7 + (-3) + (-7) =

(-3) + 4 + (-8) =   (-3) + (-8) + (-5) =

Easy Peasy All-in-One Homeschool   EP Math 5/6 Workbook

LESSON 14 Subtracting Integers

A. You have learned in the previous lesson how to add integers. Let's review.

$4 + (-5) = 4 - 5 = -1$ $\qquad (-2) + (-6) = -(2 + 6) = -8$
Adding -5 is the same as subtracting 5. $\qquad$ Add 2 and 6 in the negative direction.

B. Subtracting a number is the same as adding its opposite. When subtracting an integer, rewrite the subtraction as addition. Then solve the addition problem as you did above.

$3 - (-7) = 3 + 7 = 10$ $\qquad (-9) - 2 = (-9) + (-2) = -11$
Subtracting -7 is the same as adding 7. $\qquad$ Subtracting 2 is the same as adding -2.

$(-8) - (-5) = (-8) + 5 = 5 - 8 = -3$
Rewrite as an addition problem, then solve.

C. Subtract the numbers with different signs.

$2 - (-9) =$	$(-6) - 2 =$	$(-8) - (-2) =$
$7 - (-6) =$	$(-3) - 9 =$	$(-5) - (-7) =$
$5 - (-5) =$	$(-4) - 7 =$	$(-4) - (-9) =$
$1 - (-3) =$	$(-8) - 5 =$	$(-7) - (-4) =$
$4 - (-1) =$	$(-2) - 6 =$	$(-3) - (-8) =$

D. Add and subtract the numbers with different signs.

$(-2) + 7 =$	$8 - (-8) =$	$(-7) - (-4) =$
$(-5) - 5 =$	$(-6) - 3 =$	$(-2) - (-5) =$
$8 - (-4) =$	$(-4) + 4 =$	$(-3) + (-3) =$
$(-5) + 0 =$	$9 + (-6) =$	$(-8) + (-2) =$
$9 + (-5) =$	$5 - (-7) =$	$(-5) - (-8) =$
$(-3) - 8 =$	$7 + (-8) =$	$(-2) + (-9) =$

Easy Peasy All-in-One Homeschool $\qquad$ EP Math 5/6 Workbook

LESSON 14 Practice

A. Add the numbers with different signs.

8 + (-6) =	(-7) + 7 =	(-6) + (-9) =
5 + (-7) =	(-1) + 9 =	(-5) + (-4) =
8 + (-9) =	(-2) + 6 =	(-4) + (-3) =
6 + (-6) =	(-9) + 4 =	(-8) + (-2) =
2 + (-5) =	(-3) + 7 =	(-3) + (-6) =

B. Subtract the numbers with different signs.

7 − (-6) =	(-4) − 6 =	(-9) − (-9) =
2 − (-9) =	(-8) − 8 =	(-7) − (-4) =
5 − (-5) =	(-5) − 3 =	(-4) − (-2) =
8 − (-4) =	(-4) − 4 =	(-3) − (-3) =
0 − (-5) =	(-7) − 9 =	(-7) − (-2) =

C. Add and subtract the numbers with different signs.

7 + (-9) =	6 − (-3) =	(-2) − (-7) =
(-5) − 6 =	(-2) + 9 =	(-8) + (-5) =
(-9) + 6 =	(-4) − 7 =	(-9) − (-4) =
3 + (-7) =	(-2) + 4 =	(-7) + (-7) =
(-4) − 6 =	(-6) + 5 =	(-4) − (-8) =
(-2) + 8 =	3 − (-9) =	(-8) + (-9) =

Easy Peasy All-in-One Homeschool · EP Math 5/6 Workbook

LESSON 15 Word Problems: Integers

A. Here are some real word examples of negative numbers. Choose the correct answers.

A thermometer shows temperatures above zero as positive numbers and temperatures below zero as negative numbers. How would 5 degrees Celsius below zero be shown on a thermometer?

a. 5 °C
b. -5 °C

Geographers use positive numbers to represent elevation above sea level and negative numbers to represent elevation below sea level. If a submarine dives to 30 feet below sea level, what will be its elevation?

a. 30 feet
b. -30 feet

A company uses positive numbers to record earnings and negative numbers to record spending in its bookkeeping. How would a company represent earnings of $5,000?

a. $5,000
b. -$5,000

Banks use positive balances to represent that the bank owes the customer money and negative balances to represent that the customer owes the bank money. If you deposit $50 and later withdraw $35, how would the bank represent your balance?

a. $15
b. -$15

Credit card companies use negative balances to represent that the customer owes the company money. If a customer spends $100 with the credit card, how would the company represent his or her balance?

a. $100
b. -$100

B. Solve the word problems.

The temperature yesterday was 5 °C. Today it was 12 degrees colder. What was the temperature today? _____

California is 59 feet below sea level. Mt. Everest, the highest mountain on Earth, is 29,028 feet above sea level. What is the difference in elevation between California and Mt. Everest? _____

Stacy had a balance of $248 in her bank account. She wrote a check for $56, deposited a check for $135, and withdrew $210 in cash. What is her final balance? _____

Leah has a lemonade stand. Last week she spent $58 and earned $55. This week she spent $35 and earned $60. How much profit did Leah make for two weeks? _____

Easy Peasy All-in-One Homeschool — EP Math 5/6 Workbook

LESSON 15 Practice

A. Express each value as a positive or negative quantity.

Today's temperature is 68 degrees Fahrenheit above zero.

A shark is swimming 1,000 feet below sea level.

A local store had a profit of $1,200 last month.

Ron lost 5 pounds after going on a diet.

The number of children in a book club has increased by 18.

B. Solve the word problems.

The temperature in the morning was 8 °C, but it dropped to −2 °C in the evening. How much did it drop during the day?

A submarine was at −760 feet, then it ascended 235 feet. What is its new depth?

Jenny had a balance of $67 in her bank account. She withdrew $14 in cash. What is her new balance?

It is −7 °C in New York and 18 °C in San Francisco. What is the difference in temperature between the two cities?

A submarine was 320 feet below sea level, then it dove another 150 feet. What is its new depth?

The temperature in the morning was −4 °C. By 3 p.m. it had risen by 8 °C. What was the temperature at 3 p.m.?

Mia had $135 in her bank account. She withdrew $58, wrote a check for $25, and deposited a check for $80. What is her final balance?

An airplane is 1,150 feet above sea level. A submarine is 520 feet below sea level. What is the distance between the airplane and the submarine?

LESSON 16 Adding and Subtracting Integers

A. It's going to be a good habit for you to rewrite questions. If the signs are the same, then you find the sum of the numbers. If the signs are *different*, then you find the *difference* between the numbers. Pay attention! Is the answer going to be positive or negative?

$(-4) + 7 = 7 - 4 = 3$
Signs are opposite. Find the difference.

$(-4) + (-7) = -(4 + 7) = -11$
Signs are the same. Find the sum.
The sum is negative.

$4 - (-7) = 4 + 7 = 11$
Subtracting the negative is adding.

$4 + (-7) = -(7 - 4) = -3$
Signs are opposite. Find the difference.
The larger number is negative so the answer is negative.

Or, maybe you prefer simply the way below.

$4 + (-7) = 4 - 7 = -3$

B. Add and subtract the integers. Rewrite the questions.

$(-4) + 6 = 6 - 4 =$

$0 - (-6) = 0 + 6 =$

$2 + (-9) = -(9 - 2)$ or $2 - 9 =$

$(-4) + 3 =$

$(-7) + 9 =$

$8 - (-4) =$

$(-4) + 4 =$

$(-5) - 6 =$

$4 + (-7) =$

$(-9) - 7 =$

$2 - (-2) =$

$(-3) + (-2) = -(3 + 2) =$

$(-5) - (-9) = (-5) + 9 = 9 - 5 =$

$(-7) - (-1) = (-7) + 1 = -(7 - 1) =$

$(-7) - (-4) =$

$(-3) + (-3) =$

$(-1) - (-5) =$

$(-5) - (-8) =$

$(-6) - (-5) =$

$(-4) + (-9) =$

$(-2) - (-9) =$

$(-6) - (-6) =$

LESSON 16 Practice

Add and subtract with integers.

$5 + (-7) =$ $\qquad$ $9 + (-1) =$

$(-4) - (-8) =$ $\qquad$ $(-8) + (-9) =$

$6 + (-2) =$ $\qquad$ $(-4) - 7 =$

$(-5) + (-4) =$ $\qquad$ $8 + (-9) =$

$(-7) - 3 =$ $\qquad$ $4 - (-4) =$

$(-2) + 8 =$ $\qquad$ $(-9) + 6 =$

$7 - (-6) =$ $\qquad$ $(-4) + (-3) =$

$(-6) + 5 =$ $\qquad$ $0 - (-5) =$

$(-7) - (-2) =$ $\qquad$ $(-2) + 4 =$

$8 - (-4) =$ $\qquad$ $(-9) - (-9) =$

$(-7) + (-7) =$ $\qquad$ $(-7) - 8 =$

$9 - (-7) =$ $\qquad$ $6 - (-4) =$

$(-2) + (-9) =$ $\qquad$ $(-5) - 5 =$

$(-3) - (-3) =$ $\qquad$ $3 - (-9) =$

$(-4) - 6 =$ $\qquad$ $(-9) - (-4) =$

$(-7) + 8 =$ $\qquad$ $(-2) + (-5) =$

Easy Peasy All-in-One Homeschool   EP Math 5/6 Workbook

LESSON 17 Basic Multiplication

A. Multiplication is a fast way to count repeated groups of the same size. The result of a multiplication is called a **product**. Here is an example of counting an array of dots.

There are 4 rows with 9 dots in each row. What is the total number of dots? $= 4 \times 9 =$ _____

There are 9 columns with 4 dots in each column. What is the total number of dots? $= 9 \times 4 =$ _____

B. Multiplying a number by 0 means taking the number zero times, so it equals 0. Multiplying a number by 1 means taking the number once, so it equals the number itself.

$1 \times 0 =$ $0 \times 489 =$ $1 \times 3127 =$

$0 \times 0 =$ $134 \times 1 =$ $3582 \times 0 =$

C. Multiply. It is important to know your multiplication facts. Spend 5 to 10 minutes practicing multiplication facts if you have not mastered them yet.

$6 \times 9 =$ $8 \times 6 =$ $7 \times 7 =$

$5 \times 4 =$ $5 \times 7 =$ $9 \times 3 =$

$4 \times 3 =$ $8 \times 9 =$ $6 \times 11 =$

$12 \times 2 =$ $6 \times 1 =$ $7 \times 8 =$

$3 \times 6 =$ $10 \times 5 =$ $5 \times 5 =$

$9 \times 9 =$ $7 \times 6 =$ $12 \times 4 =$

$7 \times 4 =$ $6 \times 5 =$ $8 \times 8 =$

$5 \times 8 =$ $7 \times 3 =$ $3 \times 11 =$

$3 \times 10 =$ $6 \times 6 =$ $4 \times 4 =$

Easy Peasy All-in-One Homeschool   EP Math 5/6 Workbook

LESSON 17 Practice

Multiply.

3 × 7 =	3 × 4 =	9 × 4 =
11 × 2 =	5 × 6 =	8 × 7 =
4 × 4 =	3 × 5 =	11 × 9 =
6 × 7 =	5 × 5 =	5 × 4 =
9 × 3 =	4 × 3 =	4 × 6 =
5 × 8 =	8 × 6 =	8 × 3 =
8 × 4 =	9 × 10 =	2 × 6 =
7 × 2 =	6 × 5 =	7 × 5 =
6 × 8 =	4 × 7 =	10 × 2 =
9 × 7 =	8 × 5 =	9 × 6 =
10 × 6 =	11 × 4 =	4 × 8 =
2 × 8 =	3 × 6 =	9 × 9 =
6 × 3 =	6 × 9 =	10 × 7 =
9 × 5 =	3 × 9 =	7 × 4 =
7 × 3 =	8 × 8 =	8 × 9 =
5 × 9 =	6 × 6 =	7 × 9 =
3 × 3 =	8 × 2 =	0 × 7 =
9 × 8 =	7 × 6 =	4 × 5 =

Easy Peasy All-in-One Homeschool   EP Math 5/6 Workbook

LESSON 18 Multiplying by Multiples of 10

A. To multiply a number by a multiple of 10, ignore the zeros in the multiple of 10, multiply the rest of the numbers, and then add the zeros at the end of the product.

$800 \times 9 =$ __7200__ $70 \times 1200 =$ __84000__

Find 8 x 9, and add two zeros.   Find 7 x 12, and add three zeros.

B. Multiply by multiples of 10.

$7 \times 50 =$	$50 \times 8 =$	$11 \times 90 =$
$6 \times 60 =$	$60 \times 9 =$	$70 \times 11 =$
$90 \times 9 =$	$7 \times 60 =$	$12 \times 90 =$
$60 \times 4 =$	$6 \times 80 =$	$110 \times 5 =$
$80 \times 7 =$	$50 \times 9 =$	$80 \times 12 =$
$50 \times 50 =$	$40 \times 40 =$	$60 \times 120 =$
$40 \times 80 =$	$90 \times 70 =$	$110 \times 80 =$
$3 \times 700 =$	$5 \times 600 =$	$7000 \times 7 =$
$900 \times 4 =$	$800 \times 8 =$	$11 \times 600 =$
$120 \times 50 =$	$30 \times 120 =$	$4 \times 1200 =$

Easy Peasy All-in-One Homeschool   EP Math 5/6 Workbook

LESSON 18 Practice

A. Multiply by 10, 100, and 1000.

$7 \times 10 =$ $100 \times 4 =$ $10 \times 5 =$

$11 \times 10 =$ $12 \times 10 =$ $9 \times 10 =$

$100 \times 1 =$ $100 \times 6 =$ $8 \times 100 =$

$10 \times 100 =$ $1000 \times 3 =$ $0 \times 1000 =$

B. Multiply by multiples of 10.

$80 \times 5 =$ $70 \times 9 =$ $30 \times 7 =$

$90 \times 6 =$ $60 \times 5 =$ $4 \times 90 =$

$6 \times 70 =$ $5 \times 70 =$ $80 \times 8 =$

$40 \times 6 =$ $30 \times 5 =$ $90 \times 3 =$

$70 \times 80 =$ $70 \times 70 =$ $120 \times 60 =$

$50 \times 12 =$ $500 \times 5 =$ $60 \times 800 =$

$2 \times 110 =$ $9 \times 900 =$ $400 \times 40 =$

$120 \times 80 =$ $200 \times 12 =$ $90 \times 110 =$

$50 \times 110 =$ $120 \times 50 =$ $500 \times 60 =$

Easy Peasy All-in-One Homeschool   EP Math 5/6 Workbook

LESSON 19 Basic Division

A. Division is splitting a total into equal groups. Divide the total by the size of the groups to find the number of groups. Divide the total by the number of groups to find the size of the groups. Here is an example that shows the relationship between these two quantities.

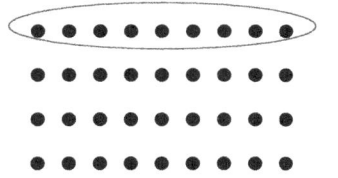

Divide 36 dots into groups of 9.
How many groups can you make? = 36 ÷ 9 = _____

Divide 36 dots into 4 equal groups.
How many dots are in each group? = 36 ÷ 4 = _____

B. Sometimes when you divide you have leftovers or **remainders**. You write the remainder after an "r" or "R" in the answer.

Divide 18 dots into groups of 4.
How many dots are remaining? = 18 ÷ 4 = _____ r _____

C. Division is the opposite of multiplication, just as addition is the opposite of subtraction.

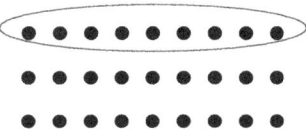

There are 4 rows of 9 dots in each row.
What is the total number of dots? = 4 × 9 = _____

Divide 36 dots into groups of 9.
How many groups can you make? = 36 ÷ 9 = _____

D. Divide by 1-digit numbers.

24 ÷ 4 =   28 ÷ 7 =   16 ÷ 2 =

30 ÷ 5 =   32 ÷ 4 =   21 ÷ 3 =

48 ÷ 6 =   63 ÷ 9 =   40 ÷ 8 =

E. Complete the division facts.

_____ ÷ 8 = 8   54 ÷ _____ = 6   27 ÷ _____ = 3

42 ÷ _____ = 7   _____ ÷ 5 = 7   _____ ÷ 9 = 9

25 ÷ _____ = 5   45 ÷ _____ = 9   20 ÷ _____ = 4

_____ ÷ 8 = 9   _____ ÷ 6 = 6   _____ ÷ 7 = 8

Easy Peasy All-in-One Homeschool   EP Math 5/6 Workbook

LESSON 19 Practice

A. Divide by 1-digit numbers.

$8 \div 4 =$ $49 \div 7 =$ $40 \div 5 =$

$36 \div 6 =$ $18 \div 3 =$ $27 \div 3 =$

$15 \div 5 =$ $63 \div 9 =$ $48 \div 4 =$

$64 \div 8 =$ $25 \div 5 =$ $35 \div 5 =$

$72 \div 9 =$ $16 \div 8 =$ $42 \div 6 =$

$21 \div 7 =$ $28 \div 4 =$ $60 \div 6 =$

B. Complete the division facts.

$9 \div \underline{\quad} = 3$ $6 \div \underline{\quad} = 2$ $\underline{\quad} \div 1 = 9$

$\underline{\quad} \div 6 = 4$ $56 \div \underline{\quad} = 7$ $\underline{\quad} \div 3 = 7$

$36 \div \underline{\quad} = 9$ $\underline{\quad} \div 9 = 5$ $54 \div \underline{\quad} = 9$

$16 \div \underline{\quad} = 8$ $63 \div \underline{\quad} = 9$ $12 \div \underline{\quad} = 2$

$\underline{\quad} \div 9 = 9$ $20 \div \underline{\quad} = 5$ $\underline{\quad} \div 3 = 8$

$14 \div \underline{\quad} = 2$ $\underline{\quad} \div 7 = 6$ $\underline{\quad} \div 8 = 9$

$\underline{\quad} \div 3 = 4$ $\underline{\quad} \div 8 = 5$ $35 \div \underline{\quad} = 5$

$\underline{\quad} \div 8 = 6$ $18 \div \underline{\quad} = 9$ $\underline{\quad} \div 8 = 4$

$10 \div \underline{\quad} = 5$ $\underline{\quad} \div 6 = 5$ $15 \div \underline{\quad} = 5$

$\underline{\quad} \div 2 = 4$ $16 \div \underline{\quad} = 4$ $\underline{\quad} \div 7 = 1$

Easy Peasy All-in-One Homeschool   EP Math 5/6 Workbook

LESSON 20 Long Division

A. Familiarize yourself with three ways to write division problems. Complete the table.

using a division symbol	using a long division symbol	as a fraction
$35 \div 5 = 7$	$5\overline{)35}$ with 7 on top	$\dfrac{35}{5} = 7$
$72 \div 8 = 9$		

B. Long division is a way to solve division problems with large numbers. It breaks down a division problem into a series of easier steps: divide, multiply, subtract, bring down, and repeat. This example shows the steps for dividing a 3-digit number by a 1-digit number.

```
      83
   _____
7 | 585
     56
    ____
     25
     21
    ____
      4
```

1. **Divide** 58 ÷ 7 = 8 r 2 and write 8 on top.
2. **Multiply** 7 x 8 = 56 and write 56 underneath 58.
3. **Subtract** 58 − 56 = 2 and write 2 underneath 56.
4. **Bring down** 5 next to 2 to make 25.
5. Repeat the steps 1 through 4 with 25.
    1) Divide 25 ÷ 7 = 3 r 4 and write 3 on top.
    2) Multiply 7 x 3 = 21 and write 21 underneath 25.
    3) Subtract 25 − 21 = 4 and write 4 underneath 21.
6. There is no digit to bring down, so 585 ÷ 7 = 83 r 4

C. Divide by 1-digit numbers. You can work out the problems in your notebook. A tip is to turn the notebook sideways and use the lines on the paper to keep the digits aligned.

$9\overline{)162}$ $7\overline{)336}$ $5\overline{)350}$ $8\overline{)912}$

$2\overline{)7744}$ $6\overline{)2964}$ $3\overline{)3639}$ $4\overline{)7336}$

Easy Peasy All-in-One Homeschool   EP Math 5/6 Workbook

LESSON 20 Practice

A. Write each division problem in three different ways.

using a division symbol	using a long division symbol	as a fraction
$21 \div 7 = 3$		
$45 \div 5 = 9$		
$32 \div 4 = 8$		

B. Divide by 1-digit numbers.

$7 \overline{)574}$ $5 \overline{)135}$ $9 \overline{)324}$ $3 \overline{)243}$

$4 \overline{)716}$ $6 \overline{)564}$ $2 \overline{)708}$ $8 \overline{)544}$

$4 \overline{)1308}$ $8 \overline{)4872}$ $2 \overline{)1158}$ $6 \overline{)1890}$

LESSON 21 Multiplying 1- and 2-Digit Numbers

A. Here are the steps for multiplying a 3-digit number by a 1-digit number with regrouping.

```
   4 6
   2 5 9
 ×     7
 ─────────
 1 8 1 3
```

1. Multiply the ones. 9 × 7 = 63
2. Write 3 in the ones column and carry 6 to the tens column.
3. Multiply the tens and add the carryover. (5 × 7) + 6 = 41
4. Write 1 in the tens column and carry 4 to the hundreds column.
5. Multiply the hundreds and add the carryover. (2 × 7) + 4 = 18
6. Write 1 in the thousands column and 8 in the hundreds column.

B. Here are the steps for multiplying a 2-digit number by a 2-digit number.

```
    1
    7
    3 8
 ×  2 9
 ───────
    3 4 2
 +  7 6 0
 ───────
  1 1 0 2
```

1. Multiply 38 × 9 and write the result in the first row.
   1) Multiply the ones. 8 × 9 = 72
   2) Write 2 in the ones column and carry 7 to the tens column.
   3) Multiply the tens and add the carryover. (3 × 9) + 7 = 34
   4) Write 4 in the tens column and 3 in the hundreds column.
2. Multiply 38 × 20 and write the result in the second row.
   1) Write 0 in the ones column.
   2) Multiply 38 × 2 as you did in the step 1.
      - 8 × 2 = 16, so write 6 and carry 1.
      - (3 × 2) + 1 = 7, so write 7.
3. Add the two products to find the answer. 342 + 760 = 1102

C. Multiply by 1-digit and 2-digit numbers.

```
      87          24         659         523         856
 ×     4     ×    9     ×     7     ×     5     ×     6
 ───────    ───────    ────────    ────────    ────────
```

```
      93          48          73          87          43
 ×    27     ×   67     ×    57     ×    83     ×    59
 ───────    ───────    ────────    ────────    ────────
```

A math riddle for you! What occurs twice in a week, once in a year, but never in a day?

Easy Peasy All-in-One Homeschool   EP Math 5/6 Workbook

LESSON 21 Practice

A. Multiply by 1-digit numbers.

$$\begin{array}{r} 876 \\ \times8 \\ \hline \end{array} \qquad \begin{array}{r} 428 \\ \times9 \\ \hline \end{array} \qquad \begin{array}{r} 367 \\ \times3 \\ \hline \end{array} \qquad \begin{array}{r} 285 \\ \times4 \\ \hline \end{array} \qquad \begin{array}{r} 579 \\ \times5 \\ \hline \end{array}$$

$$\begin{array}{r} 479 \\ \times6 \\ \hline \end{array} \qquad \begin{array}{r} 859 \\ \times3 \\ \hline \end{array} \qquad \begin{array}{r} 297 \\ \times7 \\ \hline \end{array} \qquad \begin{array}{r} 956 \\ \times5 \\ \hline \end{array} \qquad \begin{array}{r} 392 \\ \times4 \\ \hline \end{array}$$

B. Multiply the 2-digit numbers.

$$\begin{array}{r} 52 \\ \times26 \\ \hline \end{array} \qquad \begin{array}{r} 88 \\ \times37 \\ \hline \end{array} \qquad \begin{array}{r} 39 \\ \times46 \\ \hline \end{array} \qquad \begin{array}{r} 63 \\ \times35 \\ \hline \end{array} \qquad \begin{array}{r} 80 \\ \times43 \\ \hline \end{array}$$

$$\begin{array}{r} 10 \\ \times72 \\ \hline \end{array} \qquad \begin{array}{r} 48 \\ \times53 \\ \hline \end{array} \qquad \begin{array}{r} 26 \\ \times65 \\ \hline \end{array} \qquad \begin{array}{r} 29 \\ \times75 \\ \hline \end{array} \qquad \begin{array}{r} 86 \\ \times23 \\ \hline \end{array}$$

Easy Peasy All-in-One Homeschool   EP Math 5/6 Workbook

LESSON 22 Multiplying Multi-Digit Numbers

A. You can multiply any large numbers using the same technique.

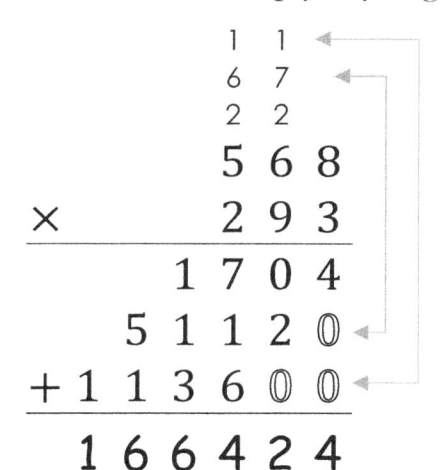

```
      1 1
      6 7
      2 2
      5 6 8
  ×     2 9 3
      1 7 0 4
      5 1 1 2 0
  + 1 1 3 6 0 0
    1 6 6 4 2 4
```

1. Multiply 568 x 3 and write the result in the first row.
2. Multiply 568 x 90 and write the result in the second row.
   1) Write 0 in the ones column.
   2) Multiply 568 x 9 as you did in the step 1.
      - 8 x 9 = 72, so write 2 and carry 7.
      - (6 x 9) + 7 = 61, so write 1 and carry 6.
      - (5 x 9) + 6 = 51, so write 51.
3. Multiply 568 x 200 and write the result in the third row.
   1) Write 0 in the tens column and ones column.
   2) Multiply 568 x 2 as you did in the step 1.
4. Add the three products to find the answer.

B. Multiply up to 4-digit numbers.

368	540	6180	2510
× 928	× 305	× 90	× 75

3975	6156	64	78
× 28	× 39	× 5476	× 7963

Easy Peasy All-in-One Homeschool   EP Math 5/6 Workbook

LESSON 22 Practice

Multiply up to 4-digit numbers.

```
    258        834         62         89         75
×     9     ×    7      ×  34      ×  73      ×  42
```

```
    283        509        329        679
×   576     ×  724     ×  615     ×  309
```

```
   4759       7125         56         39
×    60     ×   34     × 2716     × 4797
```

Easy Peasy All-in-One Homeschool   EP Math 5/6 Workbook

LESSON 23 Alternative Method: Lattice Multiplication

A. Lattice multiplication is a method of multiplying large numbers using a grid. The number of rows and columns needed in the grid depends on the number of digits in the numbers being multiplied. The example below finds 95 × 637 using lattice multiplication. The grid has 2 columns and 3 rows because there are 2 digits in 95 and 3 digits in 637.

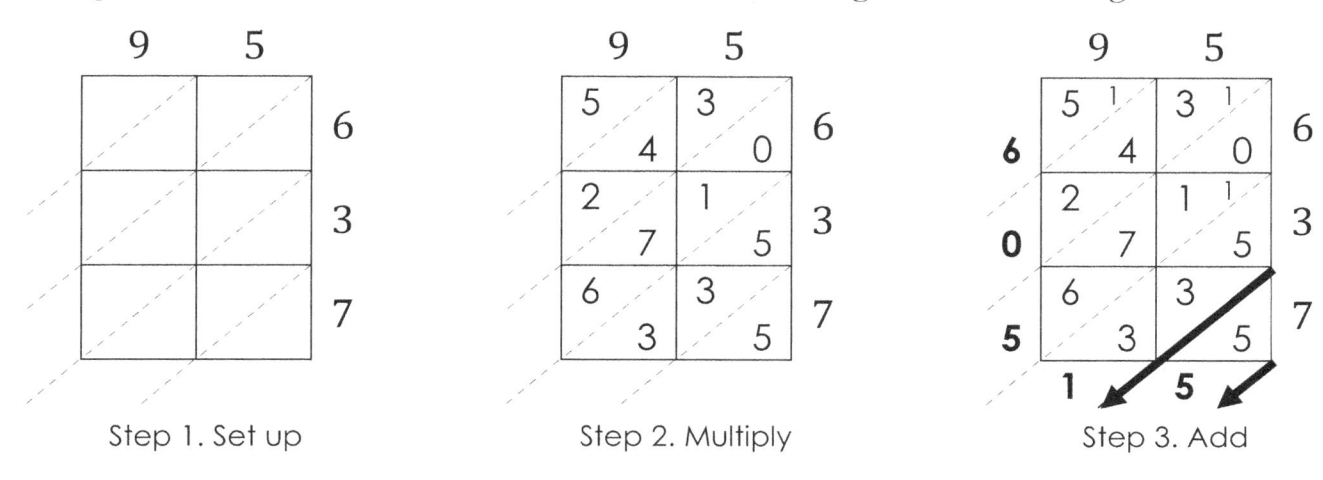

Step 1. Set up   Step 2. Multiply   Step 3. Add

1. Draw a rectangular grid matching the number of digits in the numbers you are multiplying. Draw a diagonal line in each square. Write the numbers above and to the right of the grid.
2. Multiply each row by each column and write the product in the corresponding square. Write the tens above the diagonal line and the ones below it.
3. Add up each diagonal section from right to left. Carry the tens to the next diagonal. Read the answer from top left to bottom right. The answer is 60,515.

B. Use lattice multiplication to find each product.

64 × 537 =   53 × 28 =

C. Multiply up to 4-digit numbers. Use whichever method you like.

```
    45           69           529          8517
×   93        ×  24        ×  735        ×   43
```

Easy Peasy All-in-One Homeschool   EP Math 5/6 Workbook

LESSON 23 Practice

Multiply up to 4-digit numbers. Use whichever method you like.

    786　　　　　　851　　　　　　47　　　　　　38　　　　　　62
×   7　　　　×   9　　　　× 81　　　　× 59　　　　× 47

    998　　　　　　224　　　　　　584　　　　　　290
× 727　　　× 603　　　× 322　　　× 565

  9595　　　　　8062　　　　　 75　　　　　 83
×   78　　　×   35　　　× 4239　　　× 3704

Easy Peasy All-in-One Homeschool　　　　　　　　　　EP Math 5/6 Workbook

LESSON 24 Long Division with Remainders

A. In division, the number being divided is called the **dividend**. The number dividing the dividend is called the **divisor**. The result of the division is called the **quotient**. The number left over is called the **remainder**. Identify each number in the example by its name.

$89 \div 5 = 17 \text{ r } 4$

Dividend   Divisor   Quotient   Remainder

_____ _____ _____ _____

B. You can check your answer to a division problem by using multiplication and addition. Multiply the quotient by the divisor and add the remainder. The result will be the dividend.

$89 \div 5 = 17 \text{ r } 4$ $(17 \times 5) + 4 = 89$

If you divide 89 by groups of 5, you will have 17 groups and 4 left over.   If you have 17 groups of 5 plus 4 extra, the total will be 89.

C. Check whether the answer to the division problem below is correct or not.

$218 \div 4 = 53 \text{ r } 2$

D. Find the quotients and remainders. Make sure to check your answers.

$6 \overline{)315}$ $4 \overline{)646}$ $9 \overline{)762}$ $3 \overline{)697}$

$4 \overline{)9221}$ $7 \overline{)3086}$ $5 \overline{)1638}$ $2 \overline{)3219}$

LESSON 24 Practice

Find the quotients and remainders. Make sure to check your answers.

```
     r              r              r              r
6 | 827        8 | 366        7 | 290        3 | 571


4 | 735        7 | 395        5 | 422        9 | 960


9 | 9607       3 | 8144       2 | 7641       7 | 4132


8 | 4372       8 | 2139       4 | 7245       6 | 4532
```

LESSON 25 Word Problems: Multiplying/Dividing Whole Numbers

A. When solving a word problem, it is important to read through the problem and know what is being asked. The example below shows using the strategy of working backwards.

Dylan is baking cookies for the class picnic. Fifty-six people are expected to come. He wants to prepare 3 cookies per person. A recipe calls for 1 egg to make 6 cookies. Eggs are sold in cartons of 12. How many cartons of eggs should Dylan buy?

3 cartons

Work backwards:
# of cartons = # of eggs ÷ 12
# of eggs = # of cookies ÷ 6
# of cookies = # of person × 3

Solve the problem:
# of cookies = 56 × 3 = 168
# of eggs = 168 ÷ 6 = 28
# of cartons = 28 ÷ 12 = 2 r 4

B. Solve the word problems.

There were 12 kids at Olivia's party. Olivia's Mom baked 5 pizzas and sliced each pizza into 8 pieces. Each kid ate 3 pieces. How many pieces were left?

Adam collected 43 stamps. He kept his favorite 7 stamps and then equally divided the remaining stamps among his 4 friends. How many stamps did each of his friends get?

At an apple farm, Kyle picked 18 apples. Orson picked 27 apples. Then they packed all their apples into bags. Each bag held 6 apples. How many bags did they use?

Mom bought 56 peaches. She used 28 of them to make jam. Then she baked pies with the remaining peaches. If she used 5 peaches per pie, how many peaches does she have left?

At the bookstore, Ryan bought 5 comic books that all had the same price. Joe bought 4 magazines for $12 each. Ryan spent $7 more than Joe. How much did each comic book cost?

Adam had 20 quarters, 10 dimes, and 25 nickels. He exchanged 5 quarters and 4 dimes for nickels. How many coins did Adam have in the end?

Did you know? A baseball diamond is a perfect square.

Easy Peasy All-in-One Homeschool   EP Math 5/6 Workbook

LESSON 25 Practice

Solve the word problems.

Walter has $14. His weekly allowance is $16. How many weeks will it take him to save for a video game that costs $52?

_____

A group of 27 girls and 32 boys are on a canoe trip. Each canoe can hold 7 people. How many canoes will the group need?

_____

Peter and Sam have $36 in total. Peter has three times more money than Sam. How much money does Peter have?

_____

A recipe calls for 8 apples to make one apple pie. Mom wants to bake 7 pies. Apples are sold in bags of 12. How many bags of apples does she need to buy?

_____

Ava spent $82 to order some tickets online. Each ticket cost $12. The shipping cost per order was $10. How many tickets did Ava buy?

_____

This week Kyle set a goal to spend 3 hours on math. He studied 40 minutes per day from Monday to Thursday. How many minutes does he need to study on Friday to meet his goal?

_____

Jessica had 90 beads. She made 14 bracelets with an equal number of beads and had 6 beads left. How many beads were used to make one bracelet?

_____

At an apple farm, Mia picked 15 apples. Amy picked 3 times as many apples as Mia. Then they divided all their apples equally into 4 baskets. How many apples were in each basket?

_____

Kate had 15 quarters, 15 dimes, and 15 nickels. She exchanged 4 quarters and 8 nickels for dimes. How many coins did Kate have in the end?

_____

Emma is planning a birthday party. She has $40. She wants to buy a cake for $15 and spend the rest to buy as many balloons as possible at $4 each. How much money will Emma have left?

_____

Easy Peasy All-in-One Homeschool   EP Math 5/6 Workbook

LESSON 26 Dividing by Multi-Digit Numbers

A. Here are the steps for dividing a 4-digit number by a 2-digit number.

```
         267
      _____
  29 | 7761
        58
        ___
        196
        174
        ___
         221
         203
         ___
          18
```

1. Divide 77 ÷ 29 = 2 r 19 and write 2 on top.
2. Multiply 29 × 2 = 58 and write 58 underneath 77.
3. Subtract 77 − 58 = 19 and write 19 underneath 58.
4. Bring down 6 next to 19 to make 196.
5. Repeat the steps 1 through 4 with 196.
   1) Divide 196 ÷ 29 = 6 r 22 and write 6 on top.
   2) Multiply 29 × 6 = 174 and write 174 underneath 196.
   3) Subtract 196 − 174 = 22 and write 22 underneath 174.
   4) Bring down 1 next to 22 to make 221.
6. Repeat the steps 1 through 4 with 221.
7. There is no digit to bring down, so 7761 ÷ 29 = 267 r 18.

B. Find the quotients and remainders. Make sure to check your answers.

```
     r              r              r               r
12 | 331       36 | 472       75 | 509       376 | 9271
```

```
17 | 6359      22 | 2062      56 | 9464      235 | 7903
```

A math riddle for you! Two fathers and two sons sat down to eat eggs for breakfast. They ate exactly three eggs, each person had an egg. Explain how.

LESSON 26 Practice

Find the quotients and remainders. Make sure to check your answers.

12 | 371   r   25 | 584   r   30 | 453   r   325 | 7459   r

35 | 683   52 | 871   43 | 799   267 | 9612

23 | 7238   56 | 8281   31 | 9548   358 | 5293

LESSON 27 Alternative Method: Partial Quotient Division

A. Partial quotient division is an alternative division method. It uses easy multiplication facts to find partial quotients, which are then added to find the quotient. Here are the steps for dividing multi-digit numbers using the partial quotient method.

```
            435         Partial
 18 | 7843            Quotients
      5400              300
      ----
      2443
      1800              100
      ----
       643
       540               30
       ---
       103
        90                5
        --
        13              435
     Remainder        Quotient
```

1. Write some easy multiplication facts.
   18 × 2 = 36; 18 × 3 = 54; 18 × 5 = 90
2. Find a partial quotient.
   1) How many 18s are in 7843? At least 300.
   2) Record 300 in the right column.
   3) Subtract 18 × 300 = 5400 from 7843.
3. Find another partial quotient.
   1) How many 18s are in 2443? At least 100.
   2) Record 100 in the right column.
   3) Subtract 18 × 100 = 1800 from 2443.
4. Repeat and continue finding partial quotients until the remainder is less than 18.
5. Add the partial quotients to find the quotient.
   300 + 100 + 30 + 5 = 435
6. Therefore, 7843 ÷ 18 = 435 r 13

B. Find the quotients and remainders. Use whichever method you like.

37 | 681   26 | 750   84 | 474   114 | 4104

12 | 9813   16 | 3760   59 | 6994   650 | 7491

LESSON 27 Practice

Find the quotients and remainders. Use whichever method you like.

34 | 262   52 | 781   73 | 423   150 | 6750

18 | 962   32 | 384   54 | 744   281 | 8837

42 | 9032   55 | 8670   62 | 4780   116 | 3584

34 | 8364   21 | 4630   47 | 6859   418 | 3625

LESSON 28 Word Problems: Arithmetic with Whole Numbers

Solve the word problems.

Mr. Kim bought a television set for $374. He had $158 left. How much money did Mr. Kim have at first?

The pet store sold 238 goldfish last week and 185 goldfish this week. It has 79 goldfish left. How many goldfish did the store have at first?

Dana can solve 12 addition problems per minute. How many addition problems can she solve in 5 minutes?

The librarian puts 400 books equally on 25 shelves. How many books are on each shelf?

Dominic has a 185-page reading assignment. He read 128 pages yesterday and 39 pages today. How many pages does Dominic have left to read?

Victoria found 115 seashells on the beach. Eric found 86 seashells. 37 of them were broken. How many unbroken seashells did they find together?

Derek spent $120 to order some tickets online. Each ticket cost $14. The shipping cost was $8. How many tickets did Derek buy?

Edith has 15 pennies. Lillian has 6 times as many pennies as Edith. Matt has 5 times as many pennies as Lillian. How many pennies does Matt have?

The candy store sold 156 candies last week. This week it brought in 200 more candies and sold 182 candies. Now it has 42 candies left. How many candies did the candy store have at first?

The library had a book donation campaign. Last month 1,136 fiction books and 523 non-fiction books were donated. This month 147 more fiction books and 78 fewer non-fiction books were donated than last month. How many more fiction books were donated than non-fiction books in all?

Easy Peasy All-in-One Homeschool   EP Math 5/6 Workbook

LESSON 28 Practice

Solve the word problems.

Brian has 55 baseball cards. Matt has 48 more baseball cards than Brian. How many baseball cards does Matt have? _____

Jack recycles 28 cans each month. How many cans does he recycle in a year? _____

The grocery store had 172 bags of potatoes. It sold 155 bags and brought in 86 more bags. How many bags of potatoes did the store have then? _____

Foster has $25. Clara has $16 more than Foster. Matt has $8 less than Foster. How much money do they have in all? _____

Max read a 180-page book. He read the same number of pages for 15 days. How many pages did he read each day? _____

There were 175 paper cups in the cabinet. Lucy took 36 cups from the cabinet. Carlos took twice as many cups as Lucy. How many paper cups were left in the cabinet? _____

Julio saved $20 in May. He saved $18 in June and $15 in July. Then Julio spent $28 during summer vacation. How much money does Julio have left? _____

Adam has $16. Eric has 3 times more Adam but only half of Abby. How much money does Abby have? _____

Kyle bought 7 pounds of candies. He paid with $50 and received $8 as change. What was the price per pound? _____

Ron collected 120 flower stamps and 116 bird stamps. He gave 35 of his flower stamps and 28 of his bird stamps to his brother. How many stamps did Ron have then? _____

A math riddle for you! I am an 8-digit number made up of the digits 1, 1, 2, 2, 3, 3, 4, and 4. The ones are separated by one digit, the twos are separated by two digits, the threes are separated by three digits, and the fours are separated by four digits. What number am I?

Easy Peasy All-in-One Homeschool   EP Math 5/6 Workbook

LESSON 29 Place Value

A. You can write a number in three different forms: in **standard form**, in **expanded form**, and in **word form**. Here is an example of a number written in three forms.

Standard form:

3,268,415

Expanded form:

3,000,000 + 200,000 + 60,000 + 8,000 + 400 + 10 + 5

Word form:

three million, two hundred sixty-eight thousand, four hundred fifteen

B. Write each number in expanded form.

32,917 =

406,932 =

8,320,509 =

C. Write each number in word form.

14,037 =

260,500 =

5,702,826 =

D. Write the place value of the underlined digit in each number.

<u>5</u>8,137 ⇨ **ten thousands**   2,145,<u>8</u>36 ⇨

37<u>4</u>,923 ⇨   16,<u>2</u>38,470 ⇨

<u>4</u>,235,617 ⇨   <u>3</u>9,130,218 ⇨

Easy Peasy All-in-One Homeschool   EP Math 5/6 Workbook

LESSON 29 Practice

A. Write each number in expanded form.

21,735 =

308,054 =

9,800,106 =

B. Write each number in word form.

23,106 =

504,035 =

4,853,928 =

C. Write the place value of the underlined digit in each number.

9<u>5</u>,157 ⇨   5,<u>6</u>73,608 ⇨

127,<u>4</u>42 ⇨   <u>3</u>8,453,002 ⇨

6,0<u>9</u>5,896 ⇨   5<u>5</u>,236,867 ⇨

D. Given the digits from 2 to 8, make 7-digit numbers using each digit only once.

✓ What is the smallest number you can make? _____

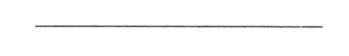

✓ What is the largest odd number you can make? _____

✓ What is the smallest number you can make that ends in 5? _____

✓ What is the largest odd number you can make that starts in 3? _____

Easy Peasy All-in-One Homeschool   EP Math 5/6 Workbook

LESSON 30 Rounding Whole Numbers

A. Rounding is used when we don't need an exact value and an estimate is good enough. To round a number to a given place value, look at the digit to the right of the rounding place. If the digit to the right is 4 or lower, round down by replacing all digits to the right of the rounding place with zeros. If the digit to the right is 5 or greater, round up by adding 1 to the rounding place digit and replacing all digits to the right of it with zeros. Here is an example of rounding a number to the nearest ten, hundred, and thousand.

	Nearest 10	Nearest 100	Nearest 1000
4,752	4,750	4,800	5,000
	The ones digit is 2. Round down!	The tens digit is 5. Round up!	The hundreds digit is 7. Round up!

B. Round each number to the nearest ten, hundred, and thousand.

	Nearest 10	Nearest 100	Nearest 1000
273	_____	_____	_____
4,528	_____	_____	_____
8,374	_____	_____	_____
63,095	_____	_____	_____
89,541	_____	_____	_____
230,487	_____	_____	_____

C. Estimate each product by rounding the top number to the nearest thousand and the bottom number to the nearest ten and then multiplying the first digits.

```
   6238  →                    4708  →
×    75  →  × _____      ×    52  →  × _____
```
   Estimate: Estimate:

Did you know? In Chinese, the word for "4" sounds like the word for "death". That is why many hospitals in China do not have a 4th floor.

Easy Peasy All-in-One Homeschool   *EP Math 5/6 Workbook*

LESSON 30 Practice

A. Round each number to the nearest ten.

36 _____	1,758 _____	43,072 _____
893 _____	2,360 _____	36,399 _____
495 _____	8,107 _____	59,828 _____

B. Round each number to the nearest hundred.

72 _____	5,359 _____	52,938 _____
649 _____	2,956 _____	94,152 _____
917 _____	4,728 _____	27,560 _____

C. Round each number to the nearest thousand.

92 _____	2,935 _____	39,938 _____
298 _____	9,486 _____	15,720 _____
735 _____	1,577 _____	64,163 _____

D. Use the clues to solve each riddle.

I am a 2-digit number. To the nearest 10, I round to 70. My tens digit is even. The sum of my digits is 11. What number am I? _____

I am a 3-digit number. I am the smallest possible number that would round to 400 when rounded to the nearest 100. What number am I? _____

I am a 4-digit number. I am the largest possible number that would round to 8,000 when rounded to the nearest 1,000. What number am I? _____

Easy Peasy All-in-One Homeschool   EP Math 5/6 Workbook

LESSON 31 Multiplying and Dividing Integers

A. When multiplying (or dividing) numbers with different signs, ignore the signs and multiply (or divide) as usual. Then determine the sign of the product (or the quotient) using two rules: 1) two like signs become a positive sign, and 2) two unlike signs become a negative sign. Here are some examples.

9 × -4 = __-36__
Positive x Negative = Negative

-5 × -7 = _____
Negative x Negative = Positive

-56 ÷ 7 = _____
Negative ÷ Positive = Negative

-40 ÷ -8 = _____
Negative ÷ Negative = Positive

B. Multiply the numbers.

6 × -9 =   -8 × -6 =   -7 × 7 =

-5 × 4 =   -5 × 7 =   9 × 3 =

4 × -3 =   8 × -9 =   0 × -1 =

8 × 0 =   -6 × -6 =   -7 × -8 =

-2 × -9 =   4 × 7 =   -9 × 9 =

C. Divide the numbers.

8 ÷ -8 =   54 ÷ 9 =   -16 ÷ 2 =

-42 ÷ 6 =   -36 ÷ 6 =   45 ÷ -5 =

25 ÷ 5 =   40 ÷ -8 =   28 ÷ 7 =

-72 ÷ -8 =   -32 ÷ -1 =   -63 ÷ 9 =

30 ÷ -6 =   -49 ÷ 7 =   -24 ÷ -4 =

Did you know? 18 is the only number that is twice the sum of its digits.

Easy Peasy All-in-One Homeschool   EP Math 5/6 Workbook

LESSON 31 Practice

A. Multiply the numbers.

$-9 \times 9 =$ $7 \times -6 =$ $-7 \times 1 =$

$7 \times -4 =$ $-6 \times -5 =$ $-8 \times -8 =$

$-5 \times -8 =$ $7 \times 3 =$ $3 \times -5 =$

$3 \times 3 =$ $-8 \times 4 =$ $-4 \times 0 =$

-157	286	-608	935	-427
× 2	× -7	× 8	× 4	× -3

B. Divide the numbers.

$-7 \div 7 =$ $-18 \div -6 =$ $32 \div -4 =$

$48 \div -6 =$ $36 \div 4 =$ $49 \div 7 =$

$-30 \div 5 =$ $-24 \div 8 =$ $81 \div -9 =$

$-16 \div -1 =$ $12 \div -6 =$ $-20 \div -5 =$

$-8 \overline{)992}$ $-7 \overline{)-406}$ $5 \overline{)-285}$ $-9 \overline{)-405}$

Easy Peasy All-in-One Homeschool   EP Math 5/6 Workbook

LESSON 32 Understanding Exponents

A. An **exponent** is a way to represent how many times a number is multiplied by itself. The expression b^n means the number b is multiplied n times. The repeated number b is called the **base**, and the number n is called the exponent, or **power**. Here are some examples.

$2 \times 2 \times 2 = 2^3$ $4 \times 4 \times 4 \times 4 \times 4 = $ _____

$5 \times 5 \times 5 \times 5 = 5^4$ $1 \times 1 \times 1 \times 1 \times 1 \times 1 \times 1 = $ _____

B. In words, the expression b^n is read "b to the power of n," "b to the n-th power," "b to the n-th," or simply "b to the n." Special cases are b^2, which is read "the square of b" or "b squared," and b^3, which is read "the cube of b" or "b cubed." Here are some examples.

5^2 5 to the power of 2 7^3 7 to the power of 3 2^6 2 to the power of 6
    5 to the 2nd power   7 to the 3rd power   2 to the 6th power
    5 to the 2nd   7 to the 3rd   2 to the 6th
    5 to the 2   7 to the 3   2 to the 6
    5 squared   7 cubed

C. Rewrite each expression using exponents.

$9 \times 9 = $ $100 \times 100 \times 100 = $

$4 \times 4 \times 4 \times 4 = $ $9 \times 9 \times 9 \times 9 \times 9 \times 9 = $

$2 \times 2 \times 2 \times 2 \times 2 \times 2 \times 2 \times 2 \times 2 \times 2 \times 2 \times 2 = $

$37 \times 37 \times 37 \times 37 \times 37 \times 37 \times 37 \times 37 \times 37 \times 37 = $

D. Evaluating an expression means finding its value. Evaluate each expression.

$1^8 = $ $5^3 = $

$2^5 = $ $3^4 = $

$0^3 = $ $4^2 = $

$12^1 = $ $10^3 = $

Easy Peasy All-in-One Homeschool   EP Math 5/6 Workbook

LESSON 32 Practice

A. Rewrite each expression using exponents.

$3 \times 3 \times 3 =$ $\qquad$ $74 \times 74 \times 74 \times 74 =$

$165 \times 165 =$ $\qquad$ $8 \times 8 \times 8 \times 8 \times 8 \times 8 =$

$1 \times 1 \times 1 \times 1 \times 1 \times 1 \times 1 \times 1 \times 1 \times 1 \times 1 =$

$10 \times 10 \times 10 \times 10 \times 10 \times 10 \times 10 \times 10 \times 10 =$

B. Evaluate each expression.

$0^9 =$ $\qquad\qquad$ $3^3 =$

$8^2 =$ $\qquad\qquad$ $9^1 =$

$1^{17} =$ $\qquad\qquad$ $2^6 =$

$10^5 =$ $\qquad\qquad$ $20^3 =$

C. Compare the expressions using >, <, or =.

$2^5 \bigcirc 5^2$ $\qquad$ $6^3 \bigcirc 1^5$ $\qquad$ $9^2 \bigcirc 3^4$

$4^3 \bigcirc 2^6$ $\qquad$ $2^3 \bigcirc 3^3$ $\qquad$ $6^2 \bigcirc 5^3$

$7^2 \bigcirc 0^9$ $\qquad$ $8^2 \bigcirc 2^7$ $\qquad$ $10^3 \bigcirc 10^2$

A math riddle for you! Add one straight line to make this true: 5 + 5 + 5 = 550

Easy Peasy All-in-One Homeschool $\qquad$ EP Math 5/6 Workbook

LESSON 33 Commutative Law of Addition and Multiplication

A. In addition, the order of the numbers being added does not affect the sum. You'll get the same answer no matter what order you add them in. This property is called the **commutative law of addition**, stated symbolically, $a + b = b + a$. The following examples illustrate this property. Add the numbers in the order they appear.

$8 + 6 =$ $7 + 25 =$ $4 + 6 + 9 =$

$6 + 8 =$ $25 + 7 =$ $9 + 4 + 6 =$

B. The commutative law also holds for multiplication. In multiplication, the order of the numbers being multiplied does not affect the product. This property is called the **commutative law of multiplication**, stated symbolically, $a \times b = b \times a$. The following examples illustrate this property. Multiply the numbers in the order they appear.

$9 \times 8 =$ $5 \times 70 =$ $7 \times 2 \times 4 =$

$8 \times 9 =$ $70 \times 5 =$ $4 \times 2 \times 7 =$

C. Use the commutative property to rewrite each expression.

$15 + 62 =$ $48 \times 17 =$

$24 + 48 =$ $92 \times 35 =$

D. Use the commutative property to solve each problem mentally. (Hint: Reorder the numbers so that you are adding a multiple of 10, or so that you are multiplying a multiple of 10 or single digits. For example, the first expression below can be reordered as $8 + 2 + 67 = 10 + 67$ and the second expression on the right as $5 \times 2 \times 9 = 10 \times 9$.)

$8 + 67 + 2 =$ $5 \times 9 \times 2 =$

$45 + 26 + 4 =$ $8 \times 2 \times 3 =$

$17 + 47 + 23 =$ $3 \times 6 \times 5 \times 7 =$

$55 + 39 + 45 =$ $5 \times 3 \times 8 \times 2 =$

Easy Peasy All-in-One Homeschool   EP Math 5/6 Workbook

LESSON 33 Practice

A. Use the commutative property to rewrite each expression.

$88 + 23 =$ $\qquad$ $13 \times 47 =$

$63 + 52 =$ $\qquad$ $72 \times 20 =$

B. Circle all expressions that are equivalent to the first one in each row.

40 + 38	38 ÷ 40	40 x 38	38 + 40	40 − 38
6 x 7 x 3	7 x 6 x 3	3 − 7 − 6	3 + 7 + 6	3 x 6 x 7
8 + 2 + 5	2 x 5 x 8	5 + 2 + 8	8 − 5 − 2	2 + 8 + 5

C. Use the commutative property to solve each problem mentally.

$39 + 8 + 2 =$ $\qquad$ $2 \times 8 \times 4 =$

$17 + 15 + 3 =$ $\qquad$ $6 \times 3 \times 3 =$

$9 + 75 + 11 =$ $\qquad$ $5 \times 3 \times 4 \times 7 =$

$28 + 35 + 45 =$ $\qquad$ $7 \times 2 \times 5 \times 9 =$

D. Write examples that show the commutative laws of addition and multiplication.

Commutative law of addition: $\qquad$ Commutative law of multiplication:

_____   _____

_____   _____

Math terms to know! All parts of an equation have names. In addition, you add the <u>addends</u> to get the <u>sum</u>. In subtraction, you subtract the <u>subtrahend</u> from the <u>minuend</u> to find the <u>difference</u>. In multiplication, you multiply the <u>factors</u> to get the <u>product</u>. In division, you divide the <u>dividend</u> by the <u>divisor</u> to obtain the <u>quotient</u>. In equation form,

```
    Addend + Addend = Sum        Minuend − Subtrahend = Difference
    Factor × Factor = Product    Dividend ÷ Divisor = Quotient
```

Easy Peasy All-in-One Homeschool   EP Math 5/6 Workbook

LESSON 34 Associative Law of Addition and Multiplication

A. Addition and multiplication have the **associative property**, stated symbolically, (a + b) + c = a + (b + c) and (a x b) x c = a x (b x c). It means that the grouping of the numbers being added or multiplied does not affect the sum or the product. You'll get the same answer no matter where you place the parentheses. The following examples illustrate this property. Find each sum and product. Always evaluate the parentheses first!

$(9 + 4) + 6 =$ $\qquad$ $(8 \times 2) \times 4 =$

$9 + (4 + 6) =$ $\qquad$ $8 \times (2 \times 4) =$

B. Use the associative property to rewrite each expression.

$(7 + 6) + 2 =$ $\qquad$ $(6 \times 8) \times 3 =$

$3 + (8 + 5) =$ $\qquad$ $9 \times (5 \times 7) =$

C. Use the associative property to solve each problem mentally. (Hint: Place the parentheses so that you are adding a multiple of 10 or so that you are multiplying a multiple of 10 or single digits.)

$4 + 7 + 3 =$ $\qquad$ $6 \times 2 \times 4 =$

$56 + 4 + 28 =$ $\qquad$ $3 \times 8 \times 5 =$

$37 + 55 + 5 =$ $\qquad$ $9 \times 5 \times 6 =$

D. Use the commutative and associative properties to rewrite each expression and solve.

$11 + (34 + 9) = 11 + (9 + 34) = (11 + 9) + 34 = 54$

$(23 + 8) + 17 =$

$5 + (26 + 95) =$

$(4 \times 7) \times 5 =$

$3 \times (8 \times 2) =$

Easy Peasy All-in-One Homeschool $\qquad$ EP Math 5/6 Workbook

LESSON 34 Practice

A. Use the associative property to rewrite each expression.

$(5 + 7) + 3 =$ $\qquad$ $(9 \times 2) \times 4 =$

$4 + (6 + 8) =$ $\qquad$ $5 \times (4 \times 6) =$

B. Circle all equations that show the associative property of addition or multiplication.

$9 + 8 \times 6 = 9 + (8 \times 6)$ $\qquad$ $(2 \times 5) \times 5 = 2 \times (5 \times 5)$

$(3 \times 2) \times 0 = 3 \times (2 \times 0)$ $\qquad$ $(0 + 1) + 0 = 0 \times (1 \times 0)$

$(7 \times 2) + 5 = 7 \times (2 + 5)$ $\qquad$ $(5 + 1) + 7 = 5 + (1 + 7)$

C. Use the commutative and associative properties to multiply the numbers faster.

$4 \times 32 \times 5$ $\qquad$ $5 \times 24 \times 2$ $\qquad$ $6 \times 8 \times 50$

D. Use the commutative and associative properties to solve each problem mentally.

$12 + (56 + 8) =$ $\qquad$ $(6 \times 3) \times 5 =$

$(34 + 7) + 16 =$ $\qquad$ $3 \times (8 \times 2) =$

$7 + (54 + 83) =$ $\qquad$ $(5 \times 7) \times 4 =$

$(11 + 6) + 49 =$ $\qquad$ $6 \times (9 \times 5) =$

Did you know? Algebra is a branch of mathematics that uses letters to represent numbers. The symbolic forms of the commutative property **a + b = b + a** and **a x b = b x a** are good examples. In algebra, the multiplication symbol is usually omitted to avoid confusion with the letter x. So **a x b = b x a** is often written as **ab = ba**.

LESSON 35 Distributive Law of Multiplication

A. Multiplication distributes over addition, meaning that multiplying a sum by a factor is equivalent to multiplying each addend by the factor separately and adding the products. This property is called the **distributive law of multiplication**, stated symbolically, a(b + c) = ab + ac. Multiplication distributes over subtraction as well, so a(b − c) = ab − ac also holds true. The examples below illustrate this property. Complete the second example.

$$4 \times (10 + 7) = (4 \times 10) + (4 \times 7) = 40 + 28 = 68$$

Distribute 4 to 10 and 7. This form is easier to solve than 4 x 17.

$$6 \times (20 - 5) =$$

B. You can use the distributive property in reverse. This process is called factoring because you are rewriting the expression as a multiplication of factors. Here is an example.

$$56 + 24 = (8 \times 7) + (8 \times 3) = 8 \times (7 + 3)$$

Take out (or factor out) the common factor 8.

C. Use the distributive property to rewrite each expression and solve.

$$7 \times (5 + 4) =$$

$$4 \times (8 - 3) =$$

$$9 \times (10 + 6) =$$

$$8 \times (20 - 5) =$$

D. Use the distributive property in reverse to rewrite each expression in a(b + c) or a(b − c) form, where a, b, and c represent positive whole numbers.

$$56 + 24 =$$

$$30 + 42 =$$

$$81 - 45 =$$

$$28 - 16 =$$

Easy Peasy All-in-One Homeschool   EP Math 5/6 Workbook

LESSON 35 Practice

A. Use the distributive property to rewrite each expression and solve.

$4 \times (9 + 3) =$

$7 \times (6 - 4) =$

$9 \times (8 + 2) =$

$8 \times (9 - 4) =$

$2 \times (30 + 6) =$

$5 \times (10 - 2) =$

$3 \times (70 + 5) =$

$6 \times (40 - 6) =$

B. Use the distributive property in reverse to rewrite each expression in a(b + c) or a(b − c) form, where a, b, and c represent positive whole numbers.

$9 + 15 =$

$18 - 12 =$

$49 + 35 =$

$35 - 20 =$

$28 + 40 =$

$64 - 24 =$

$55 + 88 =$

$90 - 50 =$

Easy Peasy All-in-One Homeschool   EP Math 5/6 Workbook

LESSON 36 Absolute Value

A. The **absolute value** of a number is its distance from zero on a number line. The symbol for absolute value is two vertical bars | | around a number, such that the absolute value of a number b is written as |b|. Here are some examples. Answer each question.

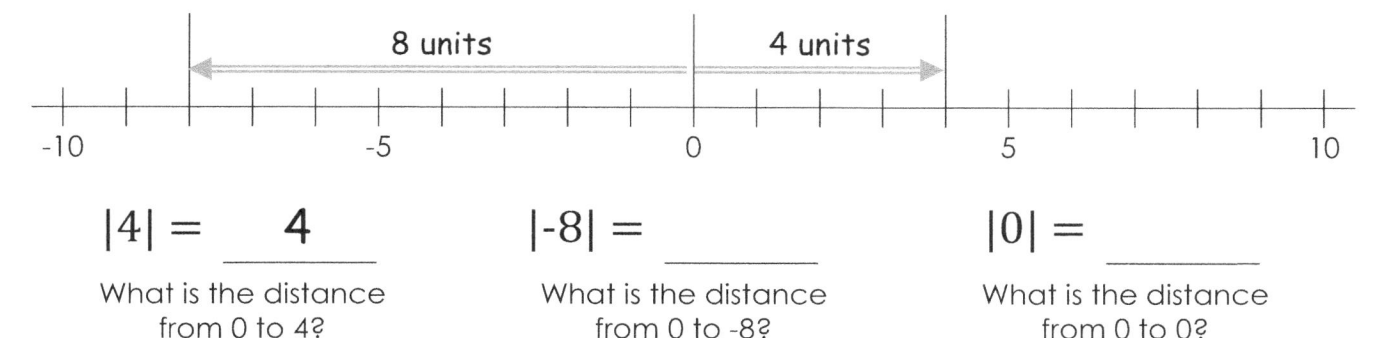

|4| = __4__
What is the distance from 0 to 4?

|-8| = _____
What is the distance from 0 to -8?

|0| = _____
What is the distance from 0 to 0?

B. Find the absolute value of each number.

|5| = |-42| = |126| =

|2| = |79| = |-413| =

|-7| = |-35| = |-528| =

C. Find the absolute value of each expression.

|5 − 8| = |6 + 27| =

|-4 + 9| = |-10 − 8| =

|23 − 7| = |-57 + -7| =

D. Circle all expressions that have an absolute value of 25. Remember to evaluate what's in the parentheses first!

5 x (2 − 7)   0 − 25   5 x 5 x 0   −10 + −15

25 ÷ 25   5 x (2 + 3)   25 x −1   5 + (2 x 3)

−5 x 5 ÷ −5   −10 + 10 + −5   5 ÷ 5 x 5   0 + −25

A math riddle for you! If 3 ⇨ 18, 4 ⇨ 32, 5 ⇨ 50, 6 ⇨ 72, 7 ⇨ 98, 8 ⇨ 128, and 9 ⇨ 162, then what is the value of 10?

Easy Peasy All-in-One Homeschool   EP Math 5/6 Workbook

LESSON 36 Practice

A. Find the absolute value of each number.

|0| = |21| = |-728| =

|-9| = |58| = |856| =

|-5| = |-99| = |-543| =

B. Find the absolute value of each expression.

|9 + -9| = |6 − 20| =

|-2 − 8| = |15 − 9| =

|7 + 34| = |-46 + -6| =

C. Circle all expressions that have an absolute value of 0. Remember to evaluate what's in the parentheses first!

0 x (1 + 1)   1 x (1 − 0)   -1 x 0   1 + (1 x 0)

1 − 2 + 1   -1 ÷ -1   -1 + 2 − 1   0 x -1000

D. Circle all expressions that have an absolute value of 1.

-1 x 1   1 − (1 x 0)   -1   1 + (1 x 1)

0 x -1   1 ÷ 1   -1 x -1   0 − 1

E. Circle all expressions that have an absolute value of 16.

-64 ÷ 8   4 x 2 x 2   8 x (2 − 4)   -4 x -4

2 x (4 + 4)   -8 − 8   4 + 4 + 4   2 + (4 x 4)

Did you know? There are different ways to write multiplication: using a times sign, an asterisk, a dot, or parentheses. For example, 7 times 6 can be written as 7 x 6, 7 * 6, 7 · 6, (7)(6), 7(6), or (7)6. However, be careful when you use a dot to indicate multiplication because it may be confused with a decimal point.

Easy Peasy All-in-One Homeschool   EP Math 5/6 Workbook

LESSON 37 Comparing Absolute Values

A. Absolute values are always positive or zero since they represent the distance between a number and zero. Here are some examples of placing absolute values on a number line.

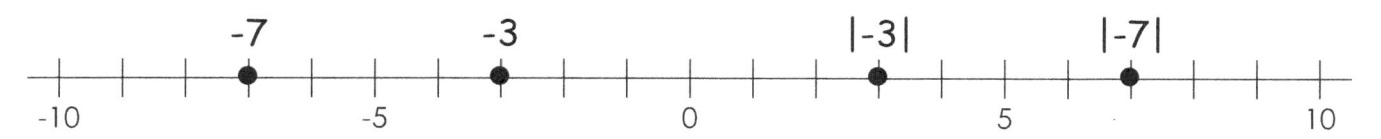

B. When numbers are marked on a number line, smaller numbers are farther to the left. Compare the numbers placed on the number line above.

-3 ◯ -7   -3 ◯ |-7| |-3| ◯ |-7|

-7 is to the left of -3.   -3 is to the left of |-7|. |-3| is to the left of |-7|.

C. Place and label the given set of numbers on the number line. Then write in the box the numbers from least to greatest.

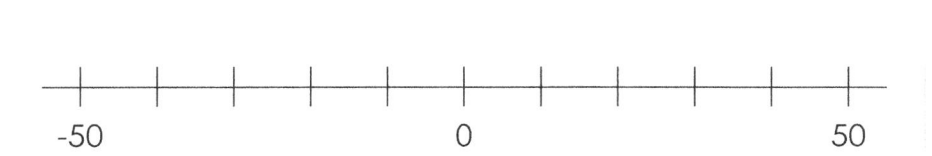

|10|, |-35|, -43, -10

D. Compare the numbers.

0 ◯ |0| 8 ◯ |-6| |3| ◯ |-7|

-6 ◯ |4| -1 ◯ |-3| |-9| ◯ |-5|

E. Order the numbers from least to greatest.

|-5|, 2, 8   10, -5, |-12| |-1|, -8, |-6|, -2

_____   _____   _____

|0|, -9, |4| -32, |-70|, 42   0, |-15|, |3|, -47

_____   _____   _____

Did you know? Forty is the only number with all its letters in alphabetical order.

Easy Peasy All-in-One Homeschool   EP Math 5/6 Workbook

LESSON 37 Practice

A. Compare the numbers.

4 ◯ -4 |4| ◯ |-4| |0| ◯ |-4|

-8 ◯ -5 |5| ◯ |-8| |-8| ◯ |-5|

-2 ◯ -9 |9| ◯ |-2| |-2| ◯ |-9|

B. Order the numbers from least to greatest.

-43, 0, -1, -82, -27, -15, -8, -30

_____

|-43|, |0|, |-1|, |-82|, |-27|, |-15|, |-8|, |-30|

_____

C. In part **B** above, the first set of numbers are all negative and the second set of numbers are the absolute values of them. What do you notice about the difference in order?

D. Order the numbers from least to greatest.

|-3|, 6, -9   12, -7, |-8| |-2|, 9, |3|, -6

_____   _____   _____

|-1|, -7, |5| -30, 22, |-14| 25, |0|, |-5|, -15

_____   _____   _____

A math riddle for you! Find three numbers whose sum and product are equal.

Easy Peasy All-in-One Homeschool   EP Math 5/6 Workbook

LESSON 38 Identity Property of 0 and 1

A. In addition, when you add 0 to any number, the sum equals the number itself. This property is called the **identity property of 0** (also called the **identity property of addition**), stated symbolically, $a + 0 = 0 + a = a$. The number 0 is called the **additive identity**. The following examples illustrate the property. Find each sum.

$1 + 0 =$ $-508 + 0 =$ $0 + 3127 =$

B. In multiplication, when you multiply any number by 1, the product equals the number itself. This property is called the **identity property of 1** (also called the **identity property of multiplication**), stated symbolically, $a \cdot 1 = 1 \cdot a = a$. The number 1 is called the **multiplicative identity**. The examples below illustrate this property. Find each product.

$1 \times 1 =$ $1 \times -508 =$ $3127 \times 1 =$

C. In multiplication, when you multiply any number by 0, the product always equals 0. This property is called the **zero product property**, stated symbolically, $a \cdot 0 = 0 \cdot a = 0$.

D. Use the number properties to solve each expression mentally. Remember to evaluate what's in the parentheses first!

$-6 \times (1 + 0) =$ $85 + (0 \times -9) =$

$-125 + |0| =$ $419 \times 27 \times 0 =$

$0 + 17 + -17 =$ $1 \times |-1| \times 35 =$

E. Evaluate each expression mentally. To evaluate expressions, replace letters (also called variables) with the given numbers. Note that multiplication symbols are omitted. For example, $539a$ means $539 \times a$. When $a = 0$, $539a = 539 \times a = 539 \times 0 = 0$.

- ✓ Evaluate $539a$ when $a = -1$.
- ✓ Evaluate $0 + a + (-8)$ when $a = 8$.
- ✓ Evaluate $41ab$ when $a = 20$ and $b = 0$.
- ✓ Evaluate $a + 76 + b$ when $a = 45$ and $b = -45$.
- ✓ Evaluate $|a| - |b|$ when $a = 100$ and $b = -100$.

Easy Peasy All-in-One Homeschool

LESSON 38 Practice

A. Write two examples that show the identity property of 0.

_____   _____

B. Write two examples that show the identity property of 1.

_____   _____

C. Write two examples that show the zero product property.

_____   _____

D. Evaluate each expression mentally.

$1{,}245 \times 1 =$ $(24 + 38) \times 0 =$

$-8 \times (1 + 0) =$ $49 + 54 + -49 =$

$|0 + -836| =$ $1 \times -1 \times 1 \times -1 =$

$-98 + 0 + 98 =$ $1{,}324 \times 0 \times 15 =$

$1 \times (512 + 0) =$ $(8 \times 1) - (8 \times 0) =$

E. Evaluate each expression mentally. Note that multiplication symbols are omitted.

- ✓ Evaluate $745a$ when $a = 1$.

- ✓ Evaluate $0 + -95 + b$ when $b = 0$.

- ✓ Evaluate $256ab$ when $a = -1$ and $b = 1$.

- ✓ Evaluate $50abc$ when $a = 3$, $b = -7$ and $c = 0$.

- ✓ Evaluate $8 + a + 12 + b$ when $a = 0$ and $b = 0$.

Easy Peasy All-in-One Homeschool   EP Math 5/6 Workbook

LESSON 39 Inverse Property of Addition and Multiplication

A. In addition, when you add any number to its opposite, the sum always equals 0. This property is called the **inverse property of addition**, stated symbolically, a + (-a) = (-a) + a = 0. The opposite of a, -a, is also called the **additive inverse** of a. Write another example that shows this property.

2,347 + (-2,347) = 0

B. In multiplication, when you multiply any number by its reciprocal, the product always equals 1. This property is called the **inverse property of multiplication**, stated symbolically, a · 1/a = 1/a · a = 1. The reciprocal of a, 1/a, is also called the **multiplicative inverse** of a. Write another example that shows this property.

$1{,}586 \times \dfrac{1}{1{,}586} = 1$

C. Use the inverse property to fill in the missing numbers.

_____ + 7 = 0 $\dfrac{1}{3}$ × _____ = 1   25 + _____ = 0

5 + _____ = 0   _____ + 2 = 0   _____ × $\dfrac{1}{25}$ = 1

_____ × 6 = 1   9 × _____ = 1   _____ + -3 = 0

$\dfrac{1}{4}$ × _____ = 1   -8 + _____ = 0   _____ × 10 = 1

D. Circle the number property that each expression illustrates.

(9 + 3) + 6 = 9 + (3 + 6)	commutative	distributive	associative
3 × (8 + 7) = (3 × 8) + (3 × 7)	distributive	associative	identity
5 + 8 + 7 = 7 + 5 + 8	commutative	distributive	identity
16 + (−16) = 0	distributive	associative	inverse
(−38) × 1 = −38	associative	identity	inverse

Easy Peasy All-in-One Homeschool   EP Math 5/6 Workbook

LESSON 39 Practice

A. Use the inverse property to fill in the missing numbers.

-5 + _____ = 0   8 + _____ = 0   9 + _____ = 0

2 × _____ = 1 $\frac{1}{9}$ × _____ = 1 $\frac{1}{40}$ × _____ = 1

_____ + 4 = 0   _____ + 6 = 0   _____ × 18 = 1

_____ × 8 = 1   _____ × $\frac{1}{12}$ = 1   _____ + -18 = 0

B. Identify the number property that each expression illustrates. Write the letter of the corresponding property in each blank. You may use a letter more than once.

A	Commutative property of addition	F	Identity property of addition
B	Commutative property of multiplication	G	Identity property of multiplication
C	Associative property of addition	H	Inverse property of addition
D	Associative property of multiplication	I	Inverse property of multiplication
E	Distributive property of multiplication	J	Zero product property

_____ 8 × (9 − 4) = (8 × 9) − (8 × 4)   _____ 2 + 4 = 4 + 2

_____ 1/6 × 6 = 1   _____ 8 + 0 = 8

_____ (3 + 2) + 7 = 3 + (2 + 7)   _____ 9 × 4 = 4 × 9

_____ 5 × 0 = 0   _____ 4 + (−4) = 0

_____ (7 × 9) × 3 = 7 × (9 × 3)   _____ 2 × 1/2 = 1

_____ 3 × (2 + 5) = (3 × 2) + (3 × 5)   _____ 8 + 9 = 9 + 8

_____ 3 × 7 = 7 × 3   _____ 3 × 1 = 3

_____ (2 × 5) × 6 = 2 × (5 × 6)   _____ (−5) + 5 = 0

LESSON 40 Dividing by Zero

A. In division, a number cannot be divided by 0. Here's why. Read each step carefully and check off its box once you understand it.

☐ Division is the opposite of multiplication. $8 \div 4 = 2$ means $4 \cdot 2 = 8$.

☐ Similarly, $8 \div 0 = b$ would mean $0 \cdot b = 8$.

☐ However, $0 \cdot b = 8$ is impossible because $0 \cdot$ anything $= 0$.

☐ There is no such **b** which, multiplied by 0, gives 8.

☐ No value works for **b**, so we say dividing by 0 is **undefined**.

B. Dividing 0 by 0 is also undefined. Here's why. Check off each box as you did above.

☐ Division is the opposite of multiplication. $8 \div 4 = 2$ means $4 \cdot 2 = 8$.

☐ Similarly, $0 \div 0 = b$ would mean $0 \cdot b = 0$.

☐ In this case, **b** can be any number because $0 \cdot$ anything $= 0$.

☐ You could say that $b = 1$ because $0 \cdot 1 = 0$, or that $b = 10$ because $0 \cdot 10 = 0$.

☐ Too many values are possible for **b**, so we say dividing 0 by 0 is **undefined**.

C. Circle all expressions that are undefined.

$0 \div 0$ 1×0 $1 \div 1$ $1 \div (1 \times 0)$

$0 \div 1$ $1 \times (0 \div 0)$ $0 \div -1$ $-1 \div 0$

$(1 \div 1) \times 0$ $1 \div 0$ $(0 \times 1) \div 0$ 0×0

D. Evaluate each expression mentally. Write "undefined" if not defined.

$75 \times 0 \times 28 =$

$45 + -45 + 45 =$

$-1 \times 94 \times -1 =$

$4^2 \div (2^3 \times 0^5) =$

$8 \times |-5| \div 0 =$

$(0 \div 5) \div (5 \div 5) =$

Easy Peasy All-in-One Homeschool

EP Math 5/6 Workbook

LESSON 40 Practice

A. Circle all expressions that are undefined.

$3 \times (0 \div 0)$ $3^3 \times 0^5 \times 1^7$ $3 \times 3 \div 3$ $|-3| \div |0|$

$0 \div |3^2|$ $(0 \div 0) - 3$ $3 \div 0^5$ $3^4 \div (-3)^2$

$3^2 \div (3^2 - 9)$ $3^3 \div (-1)^3$ $1 \div (|-3| - 3)$ $0^2 \times |-3|$

B. Write two examples for each number property.

Commutative property of addition: Commutative property of multiplication:

_____   _____

_____   _____

Associative property of addition: Associative property of multiplication:

_____   _____

_____   _____

Distributive property of multiplication: Zero product property:

_____   _____

_____   _____

Identity property of addition: Identity property of multiplication:

_____   _____

_____   _____

Inverse property of addition: Inverse property of multiplication:

_____   _____

_____   _____

Easy Peasy All-in-One Homeschool   EP Math 5/6 Workbook

LESSON 41 Arithmetic with Integers

A. Use the examples below to explain how to add and subtract numbers with different signs. These examples are from Lessons 14 and 16. Review those lessons if you need help.

-4 + 7 =   4 + -7 =

-4 + -7 =   3 − -7 =

-9 − 2 =   -8 − -5 =

B. Use the examples below to explain how to multiply and divide numbers with different signs. These examples are from Lesson 31. Review that lesson if you need help.

9 × -4 =   -5 × -7 =

-56 ÷ 7 =   -40 ÷ -8 =

C. Add, subtract, multiply, and divide the integers as indicated.

-2 + 8 =   6 − -8 =   -9 + -4 =

-9 − 3 =   3 + -6 =   9 × -3 =

-5 − -3 =   -8 − -4 =   -5 + 5 =

-3 × -6 =   6 + -8 =   9 − -7 =

25 ÷ -5 =   -21 ÷ -7 =   -4 + -7 =

2 − 6 =   5 × -8 =   -8 × -9 =

6 × -7 =   -20 ÷ -4 =   5 × -4 =

7 − -5 =   -8 + 5 =   8 + -7 =

-7 + -3 =   -49 ÷ -7 =   -64 ÷ -8 =

45 ÷ -9 =   -7 − -9 =   -8 − 2 =

Easy Peasy All-in-One Homeschool   EP Math 5/6 Workbook

LESSON 41 Practice

Add, subtract, multiply, and divide the integers as indicated.

-4 + 2 =   5 + -5 =   -3 + -6 =

-9 + -5 =   -7 + 8 =   4 + -8 =

❋ ❋ ❋ ❋ ❋ ❋ ❋ ❋ ❋

3 − -4 =   -2 − 7 =   -5 − -3 =

5 − 8 =   -9 − -7 =   8 − -6 =

❋ ❋ ❋ ❋ ❋ ❋ ❋ ❋ ❋

-2 × 9 =   6 × -6 =   -4 × -6 =

-7 × -8 =   -8 × 3 =   3 × -5 =

❋ ❋ ❋ ❋ ❋ ❋ ❋ ❋ ❋

32 ÷ -8 =   -49 ÷ 7 =   -14 ÷ -7 =

-54 ÷ -9 =   -35 ÷ 5 =   40 ÷ -5 =

❋ ❋ ❋ ❋ ❋ ❋ ❋ ❋ ❋

-9 + 4 =   -6 − -7 =   -2 + -8 =

42 ÷ -6 =   8 × -6 =   -21 ÷ 3 =

7 − -4 =   7 + -6 =   -5 × 4 =

-9 × -4 =   -27 ÷ -3 =   -6 − 6 =

4 − 9 =   -9 × 9 =   -63 ÷ 9 =

Easy Peasy All-in-One Homeschool   EP Math 5/6 Workbook

LESSON 42 Order of Operations

A. Remember how to add, subtract, multiply, and divide integers? Evaluate each expression. Review Lessons 16 and 31 if needed.

-4 + 7 =   -4 − (-7) =   9 × -4 =

B. When evaluating an expression, you may get different results depending on the order in which you perform the operations. Here is an example that demonstrates this problem.

3 + 2 × 5 = _____   3 + 2 × 5 = _____
Do addition first, then multiplication.   Do multiplication first, then addition.
What answer do you get?   What answer do you get?

C. Rules that determine the order of operations are defined to eliminate this confusion. To get correct answers, you must perform operations in the proper order: **P**arentheses, **E**xponents, **M**ultiplication and **D**ivision in order from left to right, then **A**ddition and **S**ubtraction in order from left to right. This rule is often referred to as **PEMDAS**, which you can remember as "Please Excuse My Dear Aunt Sally."

P	Parentheses
E	Exponents
MD	Multiplication & Division from LEFT to RIGHT
AS	Addition & Subtraction from LEFT to RIGHT

D. Let's first look at **MD** and **AS** in **PEMDAS**. When an expression has both multiplication and division, you solve <u>from left to right</u>, whatever comes first. You do the same with addition and subtraction. Complete the third and fourth examples.

5 × 2 × -3 ÷ -6	Multiplication	3 − 5 + 7 + -9	Subtraction
= 10 × -3 ÷ -6	Multiplication	= -2 + 7 + -9	Addition
= -30 ÷ -6	Division	= 5 + -9	Addition
= 5		= -4	

16 ÷ 4 ÷ 2 × -3	Division	7 − 5 + 3 − 8	Subtraction
=	Division	=	Addition
=	Multiplication	=	Subtraction
=		=	

(The lesson is continued on the next page. There is no practice worksheet for this lesson.)

Easy Peasy All-in-One Homeschool   EP Math 5/6 Workbook

E. Now let's look at **MDAS** in **PEMDAS**. When an expression has addition, subtraction, multiplication, and division, you first perform **MD** from left to right and then **AS** from left to right. The key idea is to take care of all multiplication and division first, before doing addition and subtraction. Complete the third and fourth examples.

$7 - 3 \times 5 + 6$	Multiplication	$3 + 2 \times {-8} \div 4 - 5$	Multiplication
$= 7 - 15 + 6$	Subtraction	$= 3 + {-16} \div 4 - 5$	Division
$= {-8} + 6$	Addition	$= 3 + {-4} - 5$	Addition
$= {-2}$		$= {-1} - 5$	Subtraction
		$= {-6}$	

$8 - 6 \times 4 \div 3$	Multiplication	$2 + 8 \div 2 - 6 \times {-3}$	Division
$=$	Division	$=$	Multiplication
$=$	Subtraction	$=$	Addition
$=$		$=$	Subtraction
		$=$	

F. Evaluate each expression using the order of operations. Do the left column first.

$3 + 2 - 7 =$

$8 \times 6 \div 3 =$

$5 + 8 \div 4 =$

$-4 \times 6 - 12 =$

$8 - 35 \div {-5} =$

$5 + {-8} - 7 - 4 + {-6} =$

$8 \times 6 \div 12 \times 4 \div {-8} =$

$3 \times 5 + {-16} \div 2 - 9 =$

$9 + 20 \div {-5} \times 4 - 3 =$

$4 \times 9 - 72 \div 9 \div {-4} =$

$5 \times 3 + 16 \div 4 - 8 \times {-2} + 6 \times 2 \times {-5} \div 10 - 8 =$

G. You will learn to evaluate expressions with parentheses and exponents in later lessons. If you are curious now, check out Lessons 52 and 138.

LESSON 43 Prime Numbers

A. A number is **prime** if it is positive and can be divided evenly (with no remainder) by only by two numbers: 1 and itself. The number 1 is *not* a prime number because it can be divided by only one number, itself. One way to identify prime numbers is to try dividing a number by all the numbers from 1 to itself. Use this straightforward method to determine whether each number is a prime number.

2 can be divided evenly by 1, 2   Is 2 a prime number? yes

4 can be divided evenly by 1, 2, 4   Is 4 a prime number? _____

5 can be divided evenly by _____   Is 5 a prime number? _____

8 can be divided evenly by _____   Is 8 a prime number? _____

9 can be divided evenly by _____   Is 9 a prime number? _____

B. Some numbers are easy to identify as not prime, because they can obviously be divided by other numbers. If a number is even, then you can divide it by 2. If a number ends in 5 or 0, then you can divide it by 5. If the sum of a number's digits is a multiple of 3, you can divide it by 3. For example, 63 is divisible by 3 because 6 + 3 = 9 is a multiple of 3. Use these rules to find all numbers that can be divided by 2, 3, or 5.

11	58	88	37	16	45	34	53
63	20	79	96	75	29	61	72
47	12	54	19	60	91	85	31

C. Write 'yes' or 'no' to indicate whether each number is a prime number.

13 _____   36 _____

55 _____   23 _____

24 _____   53 _____

31 _____   47 _____

Easy Peasy All-in-One Homeschool   EP Math 5/6 Workbook

LESSON 43 Practice

A. Circle all numbers that can be divided by 2, 3, or 5.

62   88   50   83   71   95   89   99

79   97   81   75   69   67   70   54

B. Write 'yes' or 'no' to indicate whether each number is a prime number.

16 _____   59 _____

29 _____   49 _____

72 _____   11 _____

61 _____   25 _____

37 _____   51 _____

33 _____   43 _____

C. Here are some interesting facts about prime numbers. Go tell someone about them!

- ✓ 1 is not a prime number because it can be divided by only one number, itself.
- ✓ 2 is the only even prime number. All other even numbers can be divided by 2.
- ✓ 5 is the only prime number that ends in 5. All numbers greater than 5 that end in 5 can be divided by 5.
- ✓ Every integer greater than 1 can be written as a product of prime numbers.
- ✓ Over 2,000 years ago, the Greek mathematician Euclid proved that there are infinitely many prime numbers.
- ✓ The largest known prime number has 23,249,425 digits. It was found in January 2018. Mathematicians are searching for the next prime number.

Did you know? Whole numbers are the numbers starting at 0 and counting up. Natural numbers are the whole numbers starting from 1. Integers are like whole numbers, but they also include negative numbers. Integers can be positive, negative, or zero. Zero is neither positive nor negative.

LESSON 44 Recognizing Prime Numbers

A. A number is **composite** if it is positive and can be divided evenly by one, itself, and at least one other positive integer. In other words, a composite number is a positive integer that can be formed by multiplying two smaller positive integers. Make three composite numbers out of prime numbers less than 10.

Prime x Prime = Composite   Prime x Prime = Composite

$2 \times 5 = 10$

_____   _____

_____   _____

B. Every positive integer is prime, composite, or one. One is neither prime nor composite. Write 'prime' or 'composite' for each number. If the number is composite, write at least one division fact showing it can be divided by a number other than one and itself. (Hint: Find composite numbers first! There are 6 prime numbers.)

80 composite $80 \div 2 = 40, 80 \div 5 = 16, 80 \div 8 = 10$

73 _____

97 _____

77 _____

49 _____

67 _____

83 _____

69 _____

89 _____

75 _____

31 _____

LESSON 44 Practice

A. Circle all prime numbers. (Hint: Eliminate composite numbers first! There are 25 prime numbers between 1 and 100.)

1	2	3	4	5	6	7	8	9	10
11	12	13	14	15	16	17	18	19	20
21	22	23	24	25	26	27	28	29	30
31	32	33	34	35	36	37	38	39	40
41	42	43	44	45	46	47	48	49	50
51	52	53	54	55	56	57	58	59	60
61	62	63	64	65	66	67	68	69	70
71	72	73	74	75	76	77	78	79	80
81	82	83	84	85	86	87	88	89	90
91	92	93	94	95	96	97	98	99	100

B. Where are prime numbers used in real life? Here's one example.

Prime numbers are important in Internet security. For example, when people shop online, their credit card numbers are encrypted (converted into a secure format) by a process that involves multiplying very large prime numbers. Only someone who has the secret key consisting of those prime numbers can decrypt the information. Even if an encrypted message is stolen, without the key it is nearly impossible to figure out which prime numbers were multiplied. How big are the prime numbers used for encryption? One of the most popular encryption algorithms uses more than 100-digit prime numbers!

LESSON 45 Divisibility Rules

A. When a number a can be evenly divided by another number b, we say 'a is divisible by b' and the number b is called a **divisor** or **factor** of a. A straightforward way to determine divisibility is to actually do the division and see if there is any remainder.

B. However, mathematicians have found that the numbers divisible by certain divisors follow interesting patterns. For those certain divisors, you can determine divisibility by simply checking for those patterns, without doing any actual division. Those number patterns are called **divisibility rules**. Here are common divisibility rules.

An integer is divisible by

   2 if it ends in 0, 2, 4, 6, or 8.   6 if it is divisible by both 2 and 3.
   3 if the sum of its digits is divisible by 3.   9 if the sum of its digits is divisible by 9.
   4 if its last 2 digits are divisible by 4.   10 if it ends in 0.
   5 if it ends in 0 or 5.

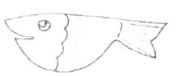

C. Determine whether each number is divisible by 2, 3, 4, 5, 6, 9, or 10. If so, list below.

48   54   80

456   570   216

6,345   9,216   1,890

25,380   45,550   83,320

Easy Peasy All-in-One Homeschool   EP Math 5/6 Workbook

LESSON 45 Practice

A. Determine whether each number is divisible by 2, 3, 4, 5, 6, 9, or 10. If so, list below.

32   45   72

240   816   140

6,410   4,425   9,000

11,133   55,590   36,594

B. Circle all numbers divisible by both 3 and 4.

531   756   592   315   304   752   888   192

396   669   103   228   702   552   257   464

C. Circle all numbers divisible by both 5 and 9.

160   990   333   954   405   855   587   729

419   585   420   180   605   234   125   270

Did you know? The mathematics and science writer Martin Gardner explained and popularized the divisibility rules in his September 1962 "Mathematical Games" column in Scientific American.

Easy Peasy All-in-One Homeschool   EP Math 5/6 Workbook

LESSON 46 Finding Factor Pairs

A. A **factor pair** is two integers that when multiplied together equal a given number. Here are the steps for finding all factor pairs of a number.

36
1 × 36
2 × 18
3 × 12
4 × 9
6 × 6
~~9 × 4~~

To find all factor pairs of 36:
1. The first pair is always 1 and the number itself. List 1 x 36.
2. Test 2. Is the number divisible by 2? If so, find the other factor that pairs with 2 and list the pair. 36 is even, so it is divisible by 2. The other factor is 36 ÷ 2 = 18. List 2 x 18.
3. Test 3 in the same way. 3 + 6 = 9, so 36 is divisible by 3. The other factor is 36 ÷ 3 = 12. List 3 x 12.
4. Continue until you have no more numbers to test. In our example, you can stop at 6 since the factors larger than 6 are already found.

B. List all factor pairs of each number.

16   55   63   21

30   88   92   70

80   56   48   64

LESSON 46 Practice

List all factor pairs of each number.

18	49	32	54

66	75	40	98

52	42	28	24

90	60	72	84

LESSON 47 Finding Factors

A. A **factor** is an integer that can evenly divide into another number. You can find all factors of a number in the same way you can find all factor pairs. The difference is that you usually list factors in order from least to greatest, not in pairs. Here are the steps for finding all factors of a number.

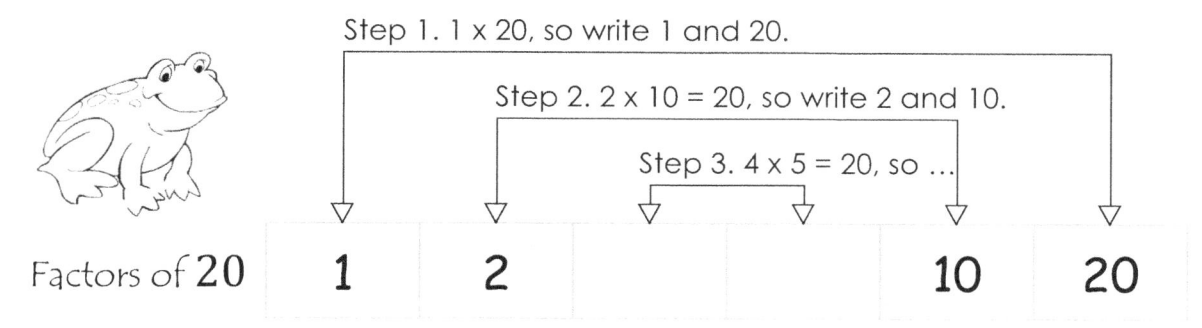

Step 1. 1 x 20, so write 1 and 20.
Step 2. 2 x 10 = 20, so write 2 and 10.
Step 3. 4 x 5 = 20, so ...

Factors of 20   1   2   10   20

B. List all factors of each number.

24 _____

63 _____

76 _____

40 _____

100 _____

140 _____

150 _____

C. Use the clues to solve each riddle.

I am a 2-digit number and a factor of 128. The sum of my digits is divisible by 10. What number am I? _____

I am a 2-digit number and a factor of 60. The product of my digits is positive. I have 4 as a factor. What number am I? _____

I am a 2-digit number and a factor of 99. The sum of my digits is divisible by 3 but not by 9. What number am I? _____

Easy Peasy All-in-One Homeschool   EP Math 5/6 Workbook

LESSON 47 Practice

A. List all factors of each number.

3 _____

6 _____

8 _____

9 _____

10 _____

56 _____

30 _____

96 _____

84 _____

72 _____

13 _____

15 _____

16 _____

18 _____

20 _____

B. Fill in the squares such that the products are correct horizontally and vertically.

2		10
3		12
6	20	

		21
		30
15	42	

		72
		28
36	56	

		64
		18
48	24	

		32
		5
4	40	

		0
		56
0	35	

Easy Peasy All-in-One Homeschool   EP Math 5/6 Workbook

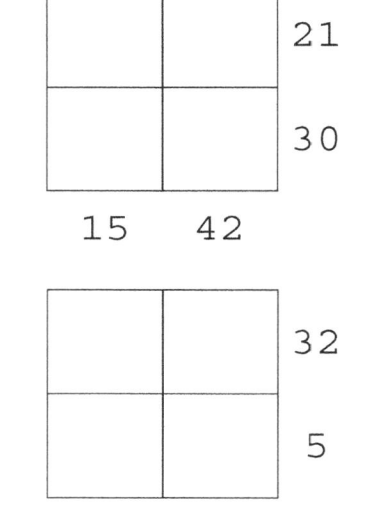

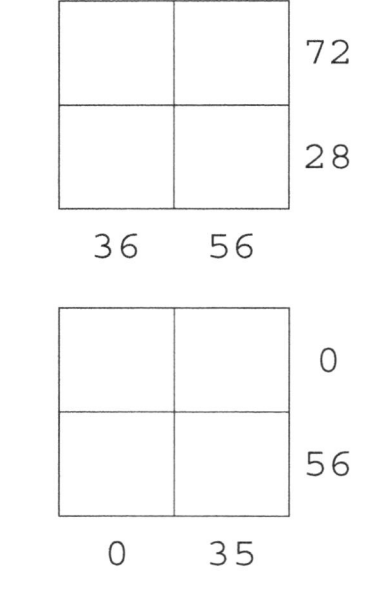

LESSON 48 Prime Factorization

A. Prime factorization is writing a number as a product of prime numbers. One way to do prime factorization is to use a **factor tree**. Here are the steps for using a factor tree to break down a number into its prime factors.

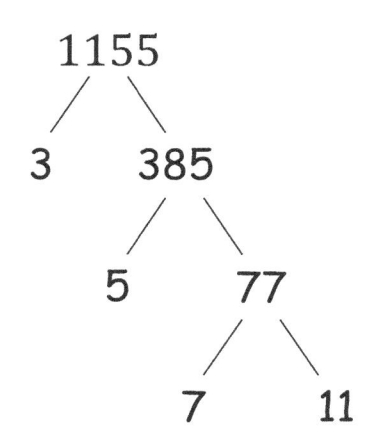

```
        1155
        /  \
       3   385
           /  \
          5    77
               /  \
              7    11
```

Therefore, the prime factorization of 1155 =

$3 \times 5 \times 7 \times 11$

To find the prime factorization of 1155:
1. Start with the smallest prime number, 2. If the number is divisible by 2, find the other factor that pairs with 2 and write them as branches of the number. 1155 is odd, so it is not divisible by 2.
2. Test the next prime number, 3, in the same way. 1 + 1 + 5 + 5 = 12 is divisible by 3. 1155 ÷ 3 = 385, so write 3 and 385 as the branches of 1155.
3. Repeat the process with 385. Test 2, 3, 5, and so on until you find a prime number that divides into 385. 385 ÷ 5 = 77, so write 5 and 77 under 385.
4. Continue breaking down factors until all the branches have prime numbers.
5. Multiply all the prime numbers at the end of each branch.

B. Find the prime factorization of each number.

70   45   88   78

525   612   126   440

LESSON 48 Practice

Find the prime factorization of each number.

35	18	52	66

44	75	85	60

546	245	375	510

924	882	294	780

LESSON 49 Least Common Multiple (LCM)

A. A **multiple** is a number you get when you multiply by an integer. The times tables are good example of multiples. All answers in the 7 times table are multiples of 7.

B. The **least common multiple (LCM)** of two integers is the smallest positive integer that is a multiple of both integers. One way to find the LCM is to list the multiples of each number until you find the first one in common. See the example below. What is the smallest multiple in common between 6 and 8? Start with the BIGGER number.

Multiples of 8: 8, 16, 24, ... continue if necessary

Multiples of 6: 6, 12, 18, 24

LCM of 6 & 8 =

C. Another method to find the LCM is to use prime factorization. This method works better than the listing method when numbers are large. Here are the steps for using the prime factorization method to find the LCM of two numbers.

$$12 = 2 \times 2 \times 3$$
$$30 = 2 \times 3 \times 5$$
$$LCM = 2 \times 2 \times 3 \times 5$$
$$= 60$$

To find the LCM of 12 and 30:
1. Write the prime factorization of each number, aligning the common factors vertically.
2. Bring down the factors in each column.
3. Multiply the factors to get the LCM.

D. Find the LCM of each number pair.

| 4 | 12 | 14 |
| 6 | 20 | 21 |

| 16 | 10 | 18 |
| 24 | 25 | 27 |

E. Solve using the LCM. (The parent's guide shows you how to set this up.)

In the United States, representatives are elected every 2 years. The president is elected every 4 years. Both were elected in 2016. When is the next year after 2016 that both will be elected? _____

Easy Peasy All-in-One Homeschool

EP Math 5/6 Workbook

LESSON 49 Practice

Find the least common multiple (LCM) of each number pair.

| 5 | 3 | 4 |
| 7 | 9 | 13 |

| 8 | 6 | 2 |
| 12 | 14 | 28 |

| 17 | 16 | 12 |
| 51 | 40 | 21 |

| 16 | 10 | 11 |
| 22 | 15 | 20 |

| 15 | 13 | 27 |
| 35 | 17 | 36 |

| 30 | 35 | 25 |
| 80 | 42 | 60 |

Easy Peasy All-in-One Homeschool   EP Math 5/6 Workbook

LESSON 50 Greatest Common Divisor (GCD)

A. A **divisor** (or factor) is a number that divides into another without a remainder. Review Lesson 47 if you need a reminder on how to find all the divisors of a number.

B. The **greatest common divisor (GCD)** of two integers is the largest positive integer that evenly divides both integers. One way to find the GCD is to list the divisors (or factors) of each number starting with the smaller number and find the greatest one in common. See the example below. What is the largest divisor in common of 24 and 36?

Divisors of 24: **1, 24, 2, 12,** ... continue if necessary

Divisors of 36: **1, 36, 2, 18, 3, 12**

1 x 36, 2 x 18, 3 x 12, 4 x 9, 6 x 6

GCD of 24 & 36 =

C. Another method to find the GCD is to use prime factorization. This method works better than the listing method when numbers are large. Here are the steps for using the prime factorization method to find the GCD of two numbers.

$$30 = 2 \times 3 \times 5$$
$$70 = 2 \times 5 \times 7$$
$$GCD = 2 \times 5$$
$$= 10$$

To find the GCD of 30 and 70:
1. Write the prime factorization of each number, aligning the common factors vertically.
2. Bring down the common factors only.
3. Multiply the factors to get the GCD.

D. Find the GCD of each number pair.

4
10

8
44

6
15

12
18

21
35

20
30

E. Solve using the GCD. (The parent's guide shows you how to set this up.)

Mia wants to cut her wires into pieces of all the same length without remainder. One wire is 52 inches long and another is 48 inches long. What is the greatest possible length of the pieces? _____

Easy Peasy All-in-One Homeschool

EP Math 5/6 Workbook

LESSON 50 Practice

Find the greatest common divisor (GCD) of each number pair.

4	2	6
6	8	9

5	9	4
20	12	18

9	7	8
33	21	36

20	42	12
35	63	56

16	14	28
52	49	70

25	16	45
30	27	75

Easy Peasy All-in-One Homeschool   EP Math 5/6 Workbook

LESSON 51 Word Problems: LCMs and GCDs

A. Explain how to find the LCM of 9 and 12. Review Lesson 49 if needed.

B. Explain how to find the GCD of 45 and 60. Review Lesson 50 if needed.

C. Word problems about LCM and GCD can be tricky. Draw pictures or use smaller numbers to understand the situation clearly. It also helps to read the problem several times. Here are two example problems. Try to solve them on your own!

Bus A runs every 9 minutes. Bus B runs every 12 minutes. Both buses leave a station at 7 a.m. Find the next time the two buses will leave the station at the same time.

> Understand the situation:
> Each bus will leave the station after:
>
> Bus A: 0, 9 min, 18 min, 27 min, 36 min, 45 min
> Bus B: 0, 12 min, 24 min, 36 min, 48 min
>
> The two buses will leave the station together after every common multiple of 9 and 12. The next time they leave together is the LCM of 9 and 12.
>
> Find the answer:
> $9 = 3 \times 3$ and $12 = 2 \times 2 \times 3$, so the LCM of 9 and 12 is 36.
> The two buses will leave the station at the same time at 7:36 a.m.

Rosa has 45 red beads and 60 blue beads. She wants to make identical bracelets using all the beads. What is the greatest number of bracelets Rosa can make?

> Understand the situation:
> Can she make 10 identical bracelets? No, because you can't divide 45 red beads evenly into 10 bracelets. What about 5 bracelets? Yes. You can divide both red beads and blue beads evenly into 5 bracelets. What about 3 bracelets? Yes! What about 9 bracelets? No! So, to have no beads left over, the number of bracelets should be a divisor of 45 and also a divisor of 60. Since we are looking for the greatest number of bracelets, we need to find the GCD of 45 and 60.
>
> Find the answer:
> $45 = 3 \times 3 \times 5$ and $60 = 2 \times 2 \times 3 \times 5$, so the GCD of 45 and 60 is 15.

(The lesson is continued on the next page. There is no practice worksheet for this lesson.)

D. Dana is making fruit baskets using 32 apples and 40 peaches. She wants to make identical baskets with no fruit left over. Answer each question.

- ✓ Can she make 5 identical baskets with no fruit left over by dividing 32 apples and 40 peaches evenly into the 5 baskets? Explain.
- ✓ Can she make 4 identical baskets with no fruit left over? Explain.
- ✓ Is the number of identical baskets she can make a common divisor or a common multiple of 32 and 40?
- ✓ If Dana wants to make as many identical baskets as possible, how many baskets can she make? Is it the LCM or the GCD of 32 and 40?

E. At a store, pencils are sold in packages of 12 and erasers in packages of 8. Asher wants to buy an equal number of pencils and erasers with the least amount of money.

- ✓ Can Asher buy 8 pencils and 8 erasers? Explain.
- ✓ Can Asher buy 48 pencils and 48 erasers? Explain.
- ✓ If Asher buys an equal number of pencils and erasers, is the number of pencils (or erasers) a common divisor or a common multiple of 12 and 8?
- ✓ If Asher wants to spend the least amount of money, how many pencils and how many erasers should he buy? Is it the LCM or the GCD of 12 and 8?

F. Solve the word problems.

Bus A runs every 15 minutes and Bus B every 18 minutes. If both buses arrive at the stop at 10:20 a.m., when will they next arrive again at the same time? _____

Jerry is making treat bags for his friends. He has 36 cookies and 24 candies, and he wants to use them all. If he wants to make as many identical bags as possible, how many bags can he make? _____

The teacher divides 25 boys and 35 girls into groups of the same size, and each group has the same number of boys and the same number of girls. What is the greatest number of groups that can be formed? _____

A store had a grand opening event. Every 50th customer received $100. Every 30th customer received free movie tickets. How many customers had to arrive before one of them received both $100 and free tickets? _____

Easy Peasy All-in-One Homeschool   EP Math 5/6 Workbook

LESSON 52 Order of Operations with Parentheses

A. Why do we need to follow the order of operations when we evaluate expressions? What does PEMDAS stand for? Review Lesson 42 if needed.

B. Evaluate each expression. Show your steps clearly in your notebook. These are the examples shown in Lesson 42. Review that lesson if needed.

$3 - 5 + 7 + {-9} =$

$5 \times 2 \times {-3} \div {-6} =$

$7 - 3 \times 5 + 6 =$

$3 + 2 \times {-8} \div 4 - 5 =$

C. When an expression has parentheses, evaluate what's in them first. Apply PEMDAS within the parentheses to get rid of them. Complete the third and fourth examples.

$5 + 2 \times (3 + 5)$	Parentheses	$9 \div 3 \times (5 - 7)$	Parentheses
$= 5 + 2 \times 8$	Multiplication	$= 9 \div 3 \times {-2}$	Division, then Multiplciation (left to right)
$= 5 + 16$	Addition	$= 3 \times {-2}$	
$= 21$		$= {-6}$	

$6 - (2 - 8) \div 3$	Parentheses	$5 \times ({-4} + 8 \div 2)$	Parentheses (D)
$=$	Division	$=$	Parentheses (A)
$=$	Subtraction	$=$	Multiplication
$=$		$=$	

D. Evaluate each expression using the order of operations. Do the left column first.

$(5 + 9) \div 2 =$

$2 \times ({-5} + {-8}) =$

$3 \times (3 - 8) =$

$(5 + 9 \div 1) - 8 \times 5 =$

$(8 + 7) \times {-3} =$

${-2} \times (3 + 9) \div 6 - 9 =$

${-15} \div (2 - 7) \times 5 - 25 \times 2 \div ({-5} - 5) - 4 \times 5 =$

Easy Peasy All-in-One Homeschool

EP Math 5/6 Workbook

LESSON 52 Practice

A. Evaluate each expression using the order of operations. Do the left column first.

$5 + 4 - 6 =$ $\qquad$ $5 \times (3 + 2) \times 4 =$

$8 - 9 - 5 =$ $\qquad$ $(7 + 9) \div 2 \times -6 =$

$5 \times 8 \div -10 =$ $\qquad$ $(3 - 6) \times (4 - 8) =$

$-81 \div 9 \div 3 =$ $\qquad$ $8 - (3 + 12) \div 5 =$

$-5 + 8 \times 5 =$ $\qquad$ $7 + 3 \times (-6 - 6) \div -3 =$

$9 - 49 \div -7 =$ $\qquad$ $(-6 + 12 \div 2 \times -4) - 5 =$

B. Here are some challenging expressions. See if you can evaluate!

$3 + 7 \times 5 + (6 + 8 \times 5 \div 4 - 8) \div 4 \times 7 =$

$27 \div (9 + -6 \times -3) \times 3 + 7 + 6 \times (5 - 8) \div 9 =$

C. Fill in the missing number in each equation.

____ $\div 3 \times 2 = 12$ $\qquad$ $3 \times 8 -$ ____ $\times 4 = 16$

$13 \times$ ____ $- 9 = 30$ $\qquad$ $-2 + 16 \div 2 +$ ____ $= 10$

Easy Peasy All-in-One Homeschool $\qquad$ EP Math 5/6 Workbook

LESSON 53 Identifying Numerators and Denominators

A. Fractions represent a part of a whole or a part of a set. They are usually written as one number over another separated by a horizontal line. The top number, called the **numerator**, represents the number of equal parts being considered. The bottom number, called the **denominator**, represents the number of equal parts in the whole. Here is an example of using a fraction to represent the shaded part of a shape. Complete the example.

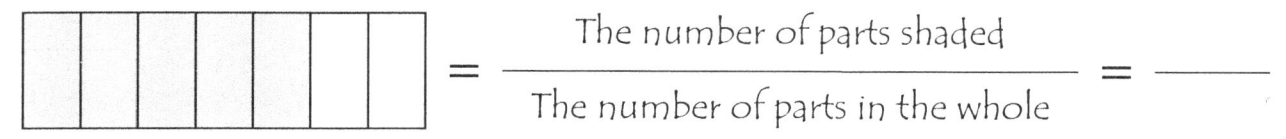

$$= \frac{\text{The number of parts shaded}}{\text{The number of parts in the whole}} = \underline{\quad\quad}$$

B. Fractions are also written as one number before another separated by a slash. In this form, the number before the slash is the numerator and the number after the slash is the denominator. Here are some fractions in different forms. Identify the numerator and denominator of each fraction.

	numerator	denominator		numerator	denominator
2/3	_____	_____	1/7	_____	_____
4/5	_____	_____	3/8	_____	_____

C. Write a fraction that represents the shaded part of each shape, then identify its numerator or denominator.

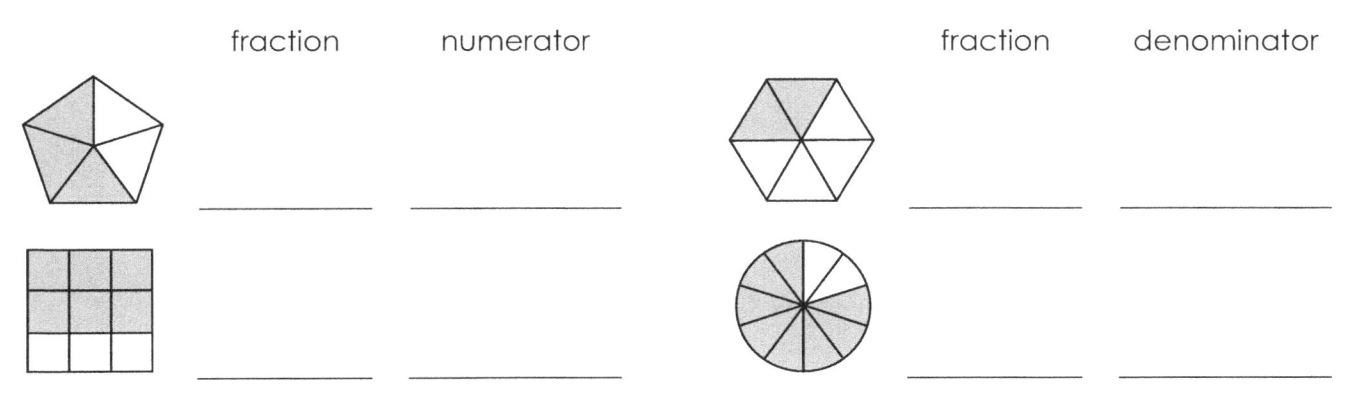

	fraction	numerator		fraction	denominator

D. What fraction am I? Use the clues to find the correct fractions.

My denominator is the greatest common divisor of 36 and 54. My numerator is the smallest 2-digit prime number. What fraction am I? _____

My numerator is the largest prime factor of 210. My denominator is the sum of all prime factors of 210. What fraction am I? _____

Easy Peasy All-in-One Homeschool   EP Math 5/6 Workbook

LESSON 53 Practice

A. Write a fraction that represents the shaded part of each shape, then identify its numerator or denominator.

fraction   numerator   fraction   denominator

_____ _____   _____ _____

_____ _____   _____ _____

_____ _____   _____ _____

B. What fraction am I? Use the clues to find the correct fractions.

My numerator is the only even prime number. My denominator is the greatest common divisor of 12 and 15. What fraction am I? _____

My denominator is the sum of the first two prime numbers. My numerator is a factor of all numbers. What fraction am I? _____

My numerator is the greatest common divisor of 15 and 25. My denominator is the largest factor of 9. What fraction am I? _____

My denominator is the least common multiple of 6 and 9. My numerator is the second prime number. What fraction am I? _____

My denominator is the least common multiple of 6 and 15. My numerator is the number of faces in a cube. What fraction am I? _____

My numerator is the largest 2-digit integer. My denominator is the smallest 3-digit positive integer. What fraction am I? _____

My denominator is the number of seconds in two minutes. My numerator is half of my denominator. What fraction am I? _____

My numerator is the largest prime factor of 195. My denominator is 2 to the fifth power, or 2^5. What fraction am I? _____

Easy Peasy All-in-One Homeschool   EP Math 5/6 Workbook

LESSON 54 Identifying Fraction Parts

A. Fractions are used in daily life to express information about wholes and parts. Read each situation below and write a fraction that describes the situation.

A pizza was cut into 8 equal pieces. Jerry ate 3 pieces of the pizza for dinner. What fraction of the pizza did Jerry eat? 3/8

Elizabeth went for a run on a 5-mile trail. She ran 3 miles and walked the rest. What fraction of the trail did Elizabeth walk? _____

Matthew solved 10 fraction problems and got 9 correct. What fraction of the problems did Matthew get correct? _____

Eighteen people were expected to come to Olivia's party. Fifteen people came. What fraction of the people missed the party? _____

There are 15 children in Paul's book club. Eight of them are boys. What fraction of the children are girls? _____

There are 7 milk chocolates and 9 dark chocolates in a box. What fraction of the chocolates in the box are dark chocolates? _____

B. When you read fractions, you first read the numerator as a whole number and then read the denominator as an ordinal number. If the numerator is greater than 1, the ordinal number becomes plural. Exceptions are the denominators 2 and 4. The denominator 2 is read "half" instead of "second." The denominator 4 is read "quarter" as well as "fourth." Here are some examples. Say each fraction out loud.

$\frac{1}{3}$ one third $\frac{2}{5}$ two fifths $\frac{1}{2}$ one half $\frac{3}{4}$ three fourths / three quarters

C. Write each fraction in numbers or in words.

two thirds =   five sixths = $\frac{3}{10}$ =

five halves =   six eighths =   1/7 =

one fourth =   four fifths = $^2/_9$ =

Easy Peasy All-in-One Homeschool   EP Math 5/6 Workbook

LESSON 54 Practice

A. Write a fraction that represents the shaded part of each shape, then identify its numerator or denominator.

fraction   numerator   fraction   denominator

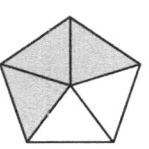

_____ _____   _____ _____

_____ _____   _____ _____

B. A circle represents one whole. Write a fraction for the shaded part of each picture.

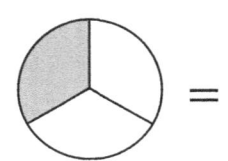

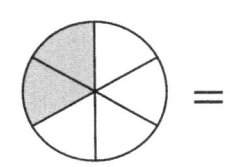

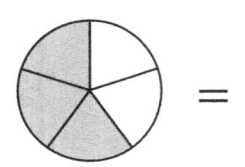

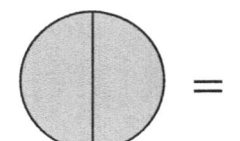

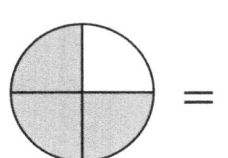

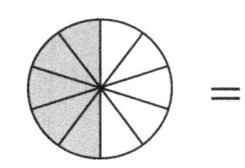

◯ =   ◯ =   ◯ =

◯ =   ◯ =   ◯ =

C. Write a fraction in numbers and in words that describes each situation.

Kyle had 2 cookies. He gave 1 cookie to his sister. What fraction of the cookies did he have left?   _____ _____

Ben had 5 gallons of paint. He used 2 gallons to paint a room. What fraction of the paint was used?   _____ _____

Andrew had $7. He spent $3 on snacks. What fraction of the money did he have <u>left</u>?   _____ _____

Emma walked 3 miles on a 4-mile trail. What fraction of the trail did she walk?   _____ _____

Amy solved 5 puzzles out of 9 puzzles. What fraction of the puzzles did Amy solve?   _____ _____

Did you know? The opposite sides of a dice always add up to seven.

LESSON 55 Equivalent Fractions

A. Fractions are **equivalent** if they represent the same amount. One way to find equivalent fractions is to use visual models (draw pictures). Here are some examples. Compare the rectangular strips and find the equal fraction in each example.

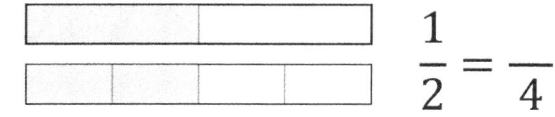

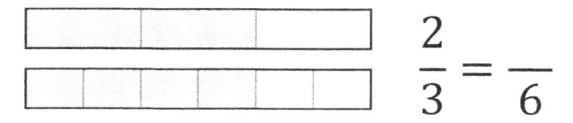

$$\frac{1}{2} = \frac{}{4} \qquad \frac{2}{3} = \frac{}{6}$$

B. Another way to find equivalent fractions is to use multiplication or division. You multiply or divide the numerator and the denominator by the same number. Here are some examples of using this method to find equal fractions. Complete the examples.

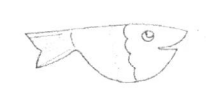

$$\frac{3}{4} = \frac{3 \times 5}{4 \times 5} = \frac{}{20} \qquad \frac{6}{18} = \frac{6 \div 6}{18 \div 6} = \frac{1}{}$$

C. Find an equivalent fraction for each fraction.

$$\frac{1}{2} = \frac{}{8} \qquad \frac{3}{7} = \frac{}{21} \qquad \frac{12}{20} = \frac{6}{} \qquad \frac{3}{4} = \frac{}{16}$$

$$\frac{18}{24} = \frac{}{4} \qquad \frac{2}{3} = \frac{6}{} \qquad \frac{3}{9} = \frac{}{72} \qquad \frac{10}{60} = \frac{1}{}$$

$$\frac{5}{6} = \frac{15}{} \qquad \frac{9}{18} = \frac{}{6} \qquad \frac{35}{56} = \frac{}{8} \qquad \frac{4}{5} = \frac{20}{}$$

D. Write at least three equivalent fractions for each fraction.

$$\frac{3}{5} \qquad\qquad\qquad \frac{20}{40}$$

$$\frac{2}{7} \qquad\qquad\qquad \frac{60}{90}$$

Easy Peasy All-in-One Homeschool   EP Math 5/6 Workbook

LESSON 55 Practice

A. Use the circle models to find equivalent fractions.

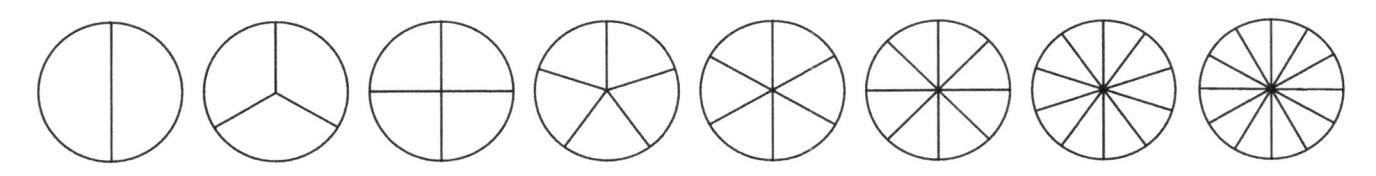

$\dfrac{3}{4} = \dfrac{}{8}$ $\dfrac{2}{6} = \dfrac{}{3}$ $\dfrac{1}{5} = \dfrac{}{10}$ $\dfrac{4}{8} = \dfrac{}{2}$

$\dfrac{2}{6} = \dfrac{}{12}$ $\dfrac{5}{10} = \dfrac{}{4}$ $\dfrac{8}{12} = \dfrac{}{3}$ $\dfrac{3}{4} = \dfrac{}{12}$

B. Use multiplication and division to find equivalent fractions.

$\dfrac{5}{15} = \dfrac{}{3}$ $\dfrac{3}{4} = \dfrac{21}{}$ $\dfrac{4}{32} = \dfrac{}{8}$ $\dfrac{3}{8} = \dfrac{}{24}$

$\dfrac{5}{8} = \dfrac{45}{}$ $\dfrac{12}{30} = \dfrac{}{5}$ $\dfrac{2}{9} = \dfrac{}{18}$ $\dfrac{5}{6} = \dfrac{25}{}$

$\dfrac{15}{20} = \dfrac{}{4}$ $\dfrac{7}{9} = \dfrac{70}{}$ $\dfrac{4}{7} = \dfrac{}{42}$ $\dfrac{35}{60} = \dfrac{}{12}$

C. What fraction am I? Use the clues to find the correct fractions.

My denominator is the number of hours in two days. I am equivalent to one half. What fraction am I? _____

My denominator is the number of pennies in one dollar. I'm equivalent to three quarters. What fraction am I? _____

My numerator is the number of sides on a rectangle. I'm equivalent to twenty hundredths. What fraction am I? _____

Easy Peasy All-in-One Homeschool   EP Math 5/6 Workbook

LESSON 56 Word Problems: Equivalent Fractions

A. To solve word problems involving fractions, first identify the relationship among the given numbers. Which number represents a whole? Which number represents part of a whole? Which fractions should be equivalent? Use this information to solve the problems. Below is an example problem. Try to solve it on your own before reading the solution.

Jenny has $50. She wants to donate one fifth of her money to charity. How many dollars should she donate? _____

> Identify the relationship among the numbers:
> A whole = The money Jenny has
> Part of a whole = The money Jenny wants to donate
> The fraction of the money Jenny wants to donate is 1/5. That is,
>
> $$\frac{\text{The amount of the money Jenny wants to donate}}{\text{The amount of the money Jenny has}} = \frac{1}{5}$$
>
> Now all we have to do is find a fraction equivalent to 1/5 whose denominator is 50. The numerator of that fraction is our answer.
>
> Find the answer:
> $\frac{?}{50} = \frac{1}{5}$, and ? = 10. Therefore, Jenny should donate $10.

B. Solve the word problems.

Ray has $36. He wants to save one fourth of his money for a trip. How many dollars should he save? _____

Sarah had $48. She spent one sixth of her money to buy a magazine. How many dollars did she spend? _____

Wendy had 39 candies. She wants to give one third of the candies to her sister. How many candies should she give? _____

Ned made a fruit basket using 16 pieces of fruit. Three quarters of the fruits were apples. How many apples were in the basket? _____

Mark has a 49-foot rope. He needs two sevenths of the rope for his art project. How many feet of the rope should he cut? _____

Leo collected 110 stamps. Two fifths of the stamps were from other countries. How many stamps were from other countries? _____

Easy Peasy All-in-One Homeschool   EP Math 5/6 Workbook

LESSON 56 Practice

A. What fraction am I? Use the clues to find the correct fractions.

I'm equivalent to 2/3. My denominator is 15. What fraction am I? _____

I'm equivalent to 1/2. My numerator is 9. What fraction am I? _____

I'm equivalent to 3/5. My numerator is 12. What fraction am I? _____

I'm equivalent to 7/3. My denominator is 33. What fraction am I? _____

I'm equivalent to 8/5. My numerator is 80. What fraction am I? _____

B. Circle all fractions that are equivalent to the first one in each row.

$$\frac{9}{15} \qquad \frac{7}{10} \qquad \frac{45}{75} \qquad \frac{18}{45} \qquad \frac{3}{5} \qquad \frac{12}{30} \qquad \frac{27}{45} \qquad \frac{45}{60} \qquad \frac{18}{30}$$

$$\frac{6}{12} \qquad \frac{3}{6} \qquad \frac{15}{21} \qquad \frac{1}{2} \qquad \frac{30}{60} \qquad \frac{12}{24} \qquad \frac{4}{9} \qquad \frac{18}{36} \qquad \frac{12}{18}$$

$$\frac{45}{60} \qquad \frac{9}{12} \qquad \frac{60}{150} \qquad \frac{3}{4} \qquad \frac{90}{120} \qquad \frac{30}{90} \qquad \frac{15}{20} \qquad \frac{20}{30} \qquad \frac{135}{180}$$

C. Solve the word problems.

A toy car is on sale. The regular price is $15, and the sale price is one third of the regular price. What is the sale price? Adam had $20, and he bought three toy cars at the sale price. What fraction of his money did he spend? _____

The math class lasts five sixths of an hour, and Danny spent three tenths of the class solving fraction problems. How many minutes long is the math class? How many minutes did Danny spent solving fraction problems? _____

Easy Peasy All-in-One Homeschool　　　　　　　　　　EP Math 5/6 Workbook

LESSON 57 Comparing Fractions

A. To compare fractions with like denominators, compare the numerators. When the denominators are the same, the fraction with the larger numerator is the larger fraction. The following examples demonstrate this visually. Compare the fractions using <, >, or =.

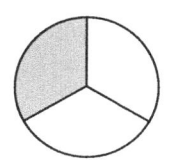

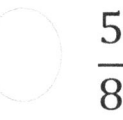

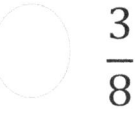

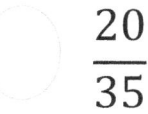

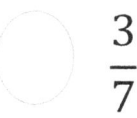

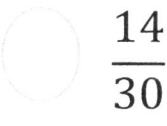

$\frac{1}{3}$ ◯ $\frac{2}{3}$ 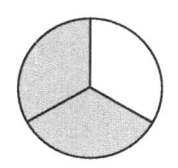 $\frac{4}{5}$ ◯ $\frac{3}{5}$

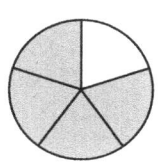

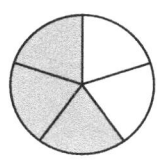

B. To compare fractions with like numerators, compare the denominators. When the numerators are the same, the fraction with the smaller denominators is the larger fraction. The following examples demonstrate this visually. Compare the fractions using <, >, or =.

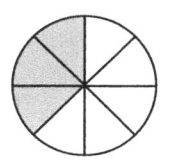

$\frac{3}{8}$ ◯ $\frac{3}{5}$ 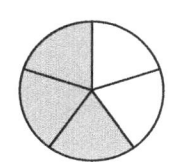 $\frac{1}{2}$ ◯ $\frac{1}{10}$

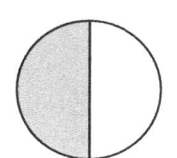

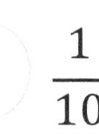

C. Compare the fractions.

$\frac{2}{3}$ ◯ $\frac{1}{3}$ $\frac{2}{6}$ ◯ $\frac{2}{9}$ $\frac{15}{21}$ ◯ $\frac{15}{14}$

$\frac{2}{4}$ ◯ $\frac{2}{8}$ $\frac{3}{8}$ ◯ $\frac{5}{8}$ $\frac{18}{16}$ ◯ $\frac{12}{16}$

$\frac{4}{9}$ ◯ $\frac{7}{9}$ $\frac{3}{2}$ ◯ $\frac{3}{8}$ $\frac{20}{35}$ ◯ $\frac{20}{36}$

$\frac{7}{8}$ ◯ $\frac{7}{9}$ $\frac{1}{4}$ ◯ $\frac{3}{4}$ $\frac{12}{35}$ ◯ $\frac{20}{35}$

$\frac{2}{3}$ ◯ $\frac{2}{5}$ $\frac{3}{8}$ ◯ $\frac{3}{7}$ $\frac{14}{28}$ ◯ $\frac{14}{30}$

LESSON 57 Practice

Compare the fractions.

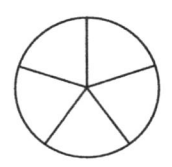

$\frac{3}{5}$ ◯ $\frac{1}{5}$  $\frac{3}{8}$ ◯ $\frac{5}{8}$

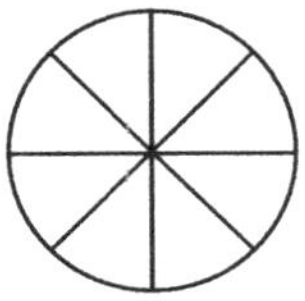

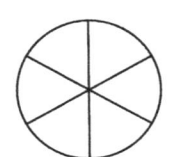

$\frac{2}{6}$ ◯ $\frac{2}{3}$  $\frac{3}{4}$ ◯ $\frac{3}{10}$

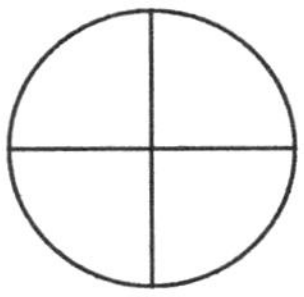

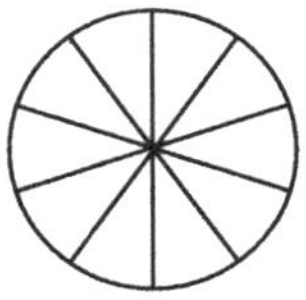

❋ ❋ ❋ ❋ ❋ ❋ ❋ ❋ ❋

$\frac{3}{5}$ ◯ $\frac{3}{8}$ $\frac{2}{6}$ ◯ $\frac{5}{6}$ $\frac{7}{13}$ ◯ $\frac{9}{13}$

$\frac{5}{7}$ ◯ $\frac{9}{7}$ $\frac{3}{4}$ ◯ $\frac{3}{5}$ $\frac{11}{15}$ ◯ $\frac{11}{45}$

$\frac{1}{8}$ ◯ $\frac{5}{8}$ $\frac{3}{8}$ ◯ $\frac{3}{4}$ $\frac{10}{23}$ ◯ $\frac{15}{23}$

$\frac{7}{9}$ ◯ $\frac{7}{8}$ $\frac{1}{6}$ ◯ $\frac{2}{6}$ $\frac{18}{30}$ ◯ $\frac{13}{30}$

$\frac{2}{4}$ ◯ $\frac{2}{7}$ $\frac{3}{9}$ ◯ $\frac{2}{9}$ $\frac{17}{15}$ ◯ $\frac{17}{30}$

Easy Peasy All-in-One Homeschool   EP Math 5/6 Workbook

LESSON 58 Comparing Fractions

A. To compare fractions with unlike denominators, make the denominators the same and then compare the numerators. One way to make the denominators the same is to use the least common multiple (LCM) as the new denominator. The following examples find the LCM by listing multiples and by using prime factorization. Complete the examples.

$\dfrac{2}{6}$? $\dfrac{4}{9}$ ⇨ Multiples of 9 = 9, **18**, 27, ...
Multiples of 6 = 12, **18**, ...
LCM = 18 ⇨ $\dfrac{}{18}$ ◯ $\dfrac{}{18}$

$\dfrac{1}{12}$? $\dfrac{3}{25}$ ⇨ 12 = 2 x 2 x 3
25 = 5 x 5
LCM = 2 x 2 x 3 x 5 x 5 ⇨ $\dfrac{}{300}$ ◯ $\dfrac{}{300}$

B. When you make the denominators the same, you can actually use any common multiple as a new denominator. The following example finds a new denominator by simply multiplying two denominators. Complete the example.

$\dfrac{2}{6}$? $\dfrac{4}{9}$ ⇨ $\dfrac{2 \times 9}{6 \times 9}$ and $\dfrac{4 \times 6}{9 \times 6}$ ⇨ $\dfrac{}{54}$ ◯ $\dfrac{}{54}$

C. Compare the fractions.

$\dfrac{1}{3}$ ◯ $\dfrac{2}{5}$ $\dfrac{4}{5}$ ◯ $\dfrac{2}{9}$ $\dfrac{8}{12}$ ◯ $\dfrac{12}{16}$

$\dfrac{4}{7}$ ◯ $\dfrac{5}{6}$ $\dfrac{3}{8}$ ◯ $\dfrac{5}{6}$ $\dfrac{7}{15}$ ◯ $\dfrac{6}{21}$

$\dfrac{1}{2}$ ◯ $\dfrac{4}{8}$ $\dfrac{2}{3}$ ◯ $\dfrac{4}{7}$ $\dfrac{5}{18}$ ◯ $\dfrac{7}{30}$

Easy Peasy All-in-One Homeschool   EP Math 5/6 Workbook

LESSON 58 Practice

A. Find an equivalent fraction for each fraction.

$\dfrac{3}{7} = \dfrac{}{14}$ $\dfrac{15}{40} = \dfrac{3}{}$ $\dfrac{5}{8} = \dfrac{}{24}$ $\dfrac{3}{4} = \dfrac{9}{}$

$\dfrac{18}{27} = \dfrac{}{9}$ $\dfrac{2}{5} = \dfrac{}{30}$ $\dfrac{20}{28} = \dfrac{10}{}$ $\dfrac{36}{63} = \dfrac{}{21}$

B. Compare the fractions.

$\dfrac{4}{5} \bigcirc \dfrac{5}{8}$ $\dfrac{3}{4} \bigcirc \dfrac{6}{8}$ $\dfrac{5}{25} \bigcirc \dfrac{3}{10}$

$\dfrac{2}{3} \bigcirc \dfrac{6}{9}$ $\dfrac{2}{5} \bigcirc \dfrac{2}{3}$ $\dfrac{6}{18} \bigcirc \dfrac{5}{15}$

$\dfrac{4}{7} \bigcirc \dfrac{1}{2}$ $\dfrac{3}{5} \bigcirc \dfrac{4}{7}$ $\dfrac{9}{14} \bigcirc \dfrac{11}{35}$

$\dfrac{2}{4} \bigcirc \dfrac{3}{6}$ $\dfrac{4}{9} \bigcirc \dfrac{3}{5}$ $\dfrac{7}{40} \bigcirc \dfrac{8}{50}$

Easy Peasy All-in-One Homeschool   EP Math 5/6 Workbook

LESSON 59 Simplifying Fractions

A. Simplifying fractions means to make the fraction as simple as possible. It is also called **reducing** since you make the numerator and denominator as small as possible without changing the value of the fraction. In other words, simplifying a fraction is finding an equivalent fraction with the smallest possible numerator and denominator. This equivalent fraction is called the simplest form, simplest terms, or lowest terms.

B. One way to simplify fractions is to divide the numerator and denominator by their greatest common divisor. Here is an example of using this method. Complete the example.

$$\frac{60}{84} \Rightarrow \begin{array}{l} 60 = \mathbf{2 \times 2 \times 3 \times 5} \\ 84 = \mathbf{2 \times 2 \times 3 \times 7} \\ GCD = 2 \times 2 \times 3 = 12 \end{array} \Rightarrow \frac{60 \div 12}{84 \div 12} = \frac{}{}$$

C. Another way to simplify fractions is to keep dividing the numerator and denominator by a common divisor until you can't divide them any further. Here is an example of reducing a fraction through this process. Complete the example.

Divide by 2 because 42 and 210 are even.

$$\frac{42}{210} = \frac{42 \div 2}{210 \div 2} = \frac{21 \div 3}{105 \div 3} = \frac{7 \div 7}{35 \div 7} = \frac{}{}$$

Divide by 3 because 2 + 1 = 3 and 1 + 5 = 6 (see Lesson 45).

D. Simplify (or reduce) each fraction to its lowest terms.

$$\frac{16}{22} = \qquad \frac{12}{16} = \qquad \frac{18}{30} =$$

$$\frac{40}{72} = \qquad \frac{36}{28} = \qquad \frac{15}{45} =$$

$$\frac{24}{108} =$$

$$\frac{315}{405} =$$

Easy Peasy All-in-One Homeschool
EP Math 5/6 Workbook

LESSON 59 Practice

Write each fraction in its simplest form.

$\dfrac{8}{18} =$ $\dfrac{10}{14} =$ $\dfrac{16}{20} =$

$\dfrac{9}{15} =$ $\dfrac{12}{27} =$ $\dfrac{35}{40} =$

$\dfrac{6}{20} =$ $\dfrac{28}{32} =$ $\dfrac{21}{28} =$

$\dfrac{9}{36} =$ $\dfrac{25}{45} =$ $\dfrac{13}{39} =$

$\dfrac{7}{42} =$ $\dfrac{14}{49} =$ $\dfrac{48}{80} =$

$\dfrac{90}{315} =$

$\dfrac{64}{200} =$

$\dfrac{140}{448} =$

Easy Peasy All-in-One Homeschool   EP Math 5/6 Workbook

LESSON 60 Ordering Fractions

A. When ordering fractions with like denominators, order them by their numerators. Here is an example. Order the fractions from least to greatest.

$$\frac{4}{11}, \frac{1}{11}, \frac{7}{11}, \frac{2}{11}, \frac{9}{11} \Rightarrow \frac{1}{11} < \frac{2}{11} < \frac{4}{11} < \frac{7}{11} < \frac{9}{11}$$

B. When ordering fractions with unlike denominators, use the least common multiple (LCM) to find equivalent fractions with like denominators. Then compare the equivalent fractions to order the original fractions. Here is an example of using this method. Follow the steps to complete the example.

First, find the LCM of the denominators using prime factorization:

$$\frac{5}{6}, \frac{3}{4}, \frac{7}{15}, \frac{9}{10}, \frac{5}{12} \Rightarrow$$

$6 = 2 \times 3$ $\quad 10 = 2 \times 5$
$4 = 2 \times 2$ $\quad 12 = 2 \times 2 \times 3$
$15 = 3 \times 5$ $\quad$ LCM $= 2 \times 2 \times 3 \times 5 = 60$

Second, find equivalent fractions and use them to order the original fractions:

$$\frac{50}{60}, \frac{}{60}, \frac{54}{60}, \frac{}{60}, \frac{}{60} \Rightarrow \frac{}{} < \frac{}{} < \frac{}{} < \frac{}{} < \frac{9}{10}$$

C. Remember that you can also find the LCM using the listing method. A quick tip is to list multiples of the largest number until you find a common multiple of the numbers. Here is how you can find the LCM of the denominators in the example above.

Order the numbers from largest to smallest: 15, 12, 10, 6, and 4. Start with the largest, 15. Is 15 also a multiple of the next largest, 12? No. Find the next multiple of 15, 15 x 2 = 30. Is 30 also a multiple of 12? No. Continue to the next multiple, 15 x 3 = 45. Is 45 also a multiple of 12? No. Continue. 15 x 4 = 60. Is 60 also a multiple of 12? Yes. Now check the other numbers. Is 60 also a multiple of 10, 6, and 4 respectively? Yes! We found the LCM.

D. Order each set of fractions from least to greatest.

$$\frac{1}{3}, \frac{2}{9}, \frac{1}{4} \qquad\qquad \frac{3}{7}, \frac{1}{3}, \frac{5}{6}$$

$$\frac{1}{2}, \frac{3}{5}, \frac{4}{10} \qquad\qquad \frac{4}{9}, \frac{7}{12}, \frac{5}{18}$$

Easy Peasy All-in-One Homeschool $\qquad\qquad$ EP Math 5/6 Workbook

LESSON 60 Practice

Order each set of fractions from least to greatest.

$\dfrac{1}{9}, \dfrac{8}{9}, \dfrac{5}{9}$ $\dfrac{2}{5}, \dfrac{2}{4}, \dfrac{2}{7}$

$\dfrac{4}{7}, \dfrac{1}{2}, \dfrac{5}{8}$ $\dfrac{4}{5}, \dfrac{1}{2}, \dfrac{1}{3}$

$\dfrac{2}{3}, \dfrac{3}{4}, \dfrac{5}{9}$ $\dfrac{3}{8}, \dfrac{5}{7}, \dfrac{1}{4}$

$\dfrac{2}{6}, \dfrac{3}{8}, \dfrac{1}{4}$ $\dfrac{4}{9}, \dfrac{2}{3}, \dfrac{5}{12}$

$\dfrac{4}{5}, \dfrac{5}{6}, \dfrac{23}{30}$ $\dfrac{3}{5}, \dfrac{13}{15}, \dfrac{11}{20}$

$\dfrac{7}{12}, \dfrac{10}{18}, \dfrac{3}{4}$ $\dfrac{7}{9}, \dfrac{11}{12}, \dfrac{13}{15}$

Easy Peasy All-in-One Homeschool   EP Math 5/6 Workbook

LESSON 61 Types of Fractions

A. There are three types of fractions: proper, improper, and mixed fractions. A **proper fraction** is a fraction where the numerator is less than the denominator. An **improper fraction** is a fraction where the numerator is greater than or equal to the denominator. A mixed fraction, or a **mixed number**, is a whole number and a proper fraction combined.

B. To read a mixed number, you use the word "and" between the whole number part and the fraction part. Here are some examples. Say each fraction out loud.

$5\frac{1}{3}$ five and one third $2\frac{3}{5}$ two and three fifths $4\frac{5}{6}$ four and five sixths $1\frac{6}{10}$ one and six tenths

C. The **reciprocal** of a fraction is the fraction turned upside down. The reciprocal of a whole number (except 0) is 1 over the number. Write the reciprocal of each fraction.

$\frac{7}{15}$ ____ $\frac{4}{9}$ ____ 15 ____ $\frac{11}{2}$ ____

D. Classify each fraction as proper, improper, or mixed. Write **P** for proper fractions, **I** for improper fractions, and **M** for mixed fractions.

$\frac{4}{11}$ ____ $\frac{14}{9}$ ____ $\frac{5}{8}$ ____ $\frac{25}{15}$ ____

$\frac{16}{16}$ ____ $8\frac{1}{2}$ ____ $4\frac{7}{9}$ ____ $\frac{29}{30}$ ____

$1\frac{2}{14}$ ____ $\frac{13}{15}$ ____ $\frac{16}{12}$ ____ $2\frac{3}{5}$ ____

E. Circle the fractions whose reciprocal is a proper fraction.

$\frac{12}{10}$ $\frac{3}{3}$ $\frac{5}{4}$ $\frac{21}{10}$ $\frac{1}{9}$ $\frac{6}{3}$ $\frac{15}{60}$ $\frac{8}{7}$

A math riddle for you! What digit is the most frequent between the numbers 1 and 1,000 (including those numbers)? What digit is the least frequent?

Easy Peasy All-in-One Homeschool   EP Math 5/6 Workbook

LESSON 61 Practice

A. Circle all proper fractions.

$\frac{4}{2}$ $\frac{2}{5}$ $\frac{8}{6}$ $\frac{11}{11}$ $\frac{5}{7}$ $\frac{3}{9}$ $\frac{15}{10}$ $\frac{1}{6}$

$\frac{12}{16}$ $\frac{7}{3}$ $\frac{20}{10}$ $\frac{2}{4}$ $\frac{30}{15}$ $\frac{4}{6}$ $\frac{8}{8}$ $\frac{9}{15}$

B. Circle all improper fractions.

$\frac{6}{5}$ $\frac{2}{3}$ $\frac{2}{10}$ $\frac{15}{9}$ $\frac{2}{8}$ $\frac{8}{7}$ $\frac{15}{15}$ $\frac{4}{6}$

$\frac{4}{2}$ $\frac{8}{8}$ $\frac{1}{5}$ $\frac{5}{7}$ $\frac{12}{6}$ $\frac{3}{4}$ $\frac{7}{10}$ $\frac{5}{3}$

C. Circle all mixed numbers.

$1\frac{7}{3}$ $4\frac{1}{2}$ $\frac{6}{8}$ $1\frac{7}{10}$ $5\frac{3}{4}$ $\frac{11}{9}$ $3\frac{2}{7}$ $2\frac{9}{8}$

$8\frac{2}{5}$ $\frac{3}{21}$ $2\frac{7}{4}$ $5\frac{3}{3}$ $6\frac{7}{9}$ $4\frac{2}{3}$ $\frac{12}{5}$ $1\frac{1}{6}$

D. Write the reciprocal of each fraction.

$\frac{36}{9}$ _____   23 _____ $\frac{10}{6}$ _____ $\frac{14}{12}$ _____

$\frac{8}{16}$ _____ $\frac{21}{7}$ _____   17 _____ $\frac{9}{42}$ _____

A math riddle for you! What is the next number in this sequence? 77, 49, 36, 18, ___

Easy Peasy All-in-One Homeschool

EP Math 5/6 Workbook

LESSON 62 Mixed Numbers to Improper Fractions

A. You can use either a mixed number or an improper fraction to show the same amount. The following example demonstrates this visually.

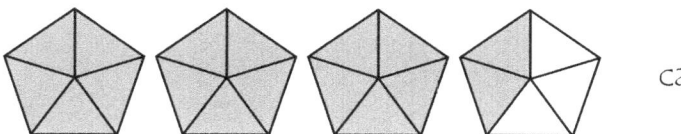

can be expressed as $3\dfrac{2}{5}$ or $\dfrac{17}{5}$

B. All mixed numbers can be written as improper fractions. Here are the steps for converting mixed numbers to improper fractions. Complete the example.

$2\dfrac{5}{6} = \dfrac{}{6}$

1. The denominator doesn't change. Keep 6 as the denominator.
2. Find the new numerator.
   1) Multiply the whole number by the denominator. 2 x 6 = 12
   2) Add that to the numerator. 12 + 5 = 17
   3) Make the result the new numerator.

C. You don't have to memorize the steps if you understand the principles behind them. Think about the following questions and compare them with the steps above.

$2\dfrac{5}{6} = \dfrac{}{6}$

1. We want to know how many sixths are in the mixed number, so the denominator 6 doesn't change.
2. How many sixths are in the whole number 2? We also have an extra 5 sixths. Then how many sixths are there in total?

D. Convert the mixed numbers to improper fractions.

$4\dfrac{1}{5} = $ $2\dfrac{3}{8} = $ $5\dfrac{2}{3} = $

$3\dfrac{4}{7} = $ $3\dfrac{1}{2} = $ $7\dfrac{5}{6} = $

$5\dfrac{3}{4} = $ $5\dfrac{4}{9} = $ $2\dfrac{3}{10} = $

Did you know? In English, every odd number contains an "e."

Easy Peasy All-in-One Homeschool   EP Math 5/6 Workbook

LESSON 62 Practice

Convert the mixed numbers to improper fractions.

$7\dfrac{1}{4} =$ $\qquad$ $4\dfrac{5}{7} =$ $\qquad$ $9\dfrac{1}{3} =$

$8\dfrac{4}{6} =$ $\qquad$ $5\dfrac{1}{2} =$ $\qquad$ $5\dfrac{7}{8} =$

$3\dfrac{7}{9} =$ $\qquad$ $1\dfrac{2}{5} =$ $\qquad$ $9\dfrac{6}{7} =$

$6\dfrac{3}{5} =$ $\qquad$ $5\dfrac{1}{10} =$ $\qquad$ $3\dfrac{7}{11} =$

$2\dfrac{1}{9} =$ $\qquad$ $3\dfrac{17}{20} =$

$4\dfrac{5}{8} =$ $\qquad$ $5\dfrac{15}{16} =$

$7\dfrac{2}{3} =$ $\qquad$ $4\dfrac{22}{25} =$

$9\dfrac{3}{4} =$ $\qquad$ $8\dfrac{11}{15} =$

Easy Peasy All-in-One Homeschool   EP Math 5/6 Workbook

LESSON 63 Improper Fractions to Mixed Numbers

A. Improper fractions can also be written as mixed numbers. Here are the steps for converting improper fractions to mixed numbers. Complete the example.

$$\frac{17}{5} = 3\frac{\ }{5}$$

1. The denominator doesn't change. Keep 5 as the denominator.
2. Find the new whole number and new numerator.
   1) Divide the numerator by the denominator. 17 ÷ 5 = 3 r 2
   2) Make the whole number quotient the new whole number.
   3) Make the remainder the new numerator.

B. As before, you don't have to memorize the steps if you understand the principles behind them. Think about the following questions and compare them with the steps above.

$$\frac{17}{5} = 3\frac{\ }{5}$$

1. We want to know how many wholes and fifths are in the improper fraction, so the denominator 5 doesn't change.
2. How many fives can be taken out of 17 with how many leftover?

C. Convert the improper fractions to mixed numbers.

$$\frac{16}{7} = \qquad \frac{13}{2} = \qquad \frac{39}{5} =$$

$$\frac{45}{6} = \qquad \frac{30}{8} = \qquad \frac{46}{3} =$$

$$\frac{27}{4} = \qquad \frac{52}{9} = \qquad \frac{37}{10} =$$

$$\frac{149}{12} =$$

$$\frac{458}{35} =$$

Easy Peasy All-in-One Homeschool   EP Math 5/6 Workbook

LESSON 63 Practice

Convert the improper fractions to mixed numbers.

$\dfrac{37}{4} =$ $\qquad$ $\dfrac{23}{2} =$ $\qquad$ $\dfrac{42}{8} =$

$\dfrac{48}{9} =$ $\qquad$ $\dfrac{51}{7} =$ $\qquad$ $\dfrac{22}{5} =$

$\dfrac{64}{3} =$ $\qquad$ $\dfrac{39}{6} =$ $\qquad$ $\dfrac{43}{4} =$

$\dfrac{26}{7} =$ $\qquad$ $\dfrac{19}{12} =$ $\qquad$ $\dfrac{30}{16} =$

$\dfrac{29}{8} =$ $\qquad$ $\dfrac{101}{15} =$

$\dfrac{17}{2} =$ $\qquad$ $\dfrac{319}{25} =$

$\dfrac{43}{5} =$ $\qquad$ $\dfrac{475}{20} =$

$\dfrac{80}{9} =$ $\qquad$ $\dfrac{232}{16} =$

Easy Peasy All-in-One Homeschool   EP Math 5/6 Workbook

LESSON 64 Comparing Mixed Numbers

A. To compare mixed numbers, compare the whole number parts first. The fraction with the larger whole number is the larger fraction. When the whole numbers are the same, compare the fraction parts. The following examples demonstrate this visually. Compare the fractions using <, >, or =.

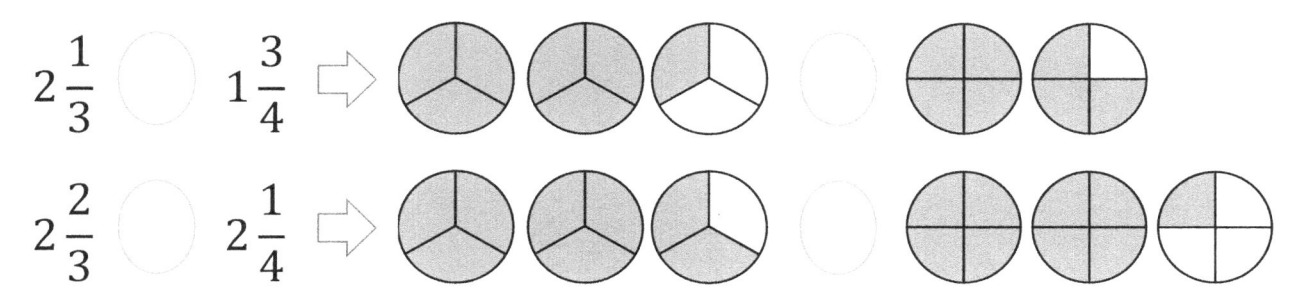

$2\frac{1}{3}$ ◯ $1\frac{3}{4}$

$2\frac{2}{3}$ ◯ $2\frac{1}{4}$

B. Another way to compare mixed numbers is to convert them to improper fractions before comparing. Here is an example of using this method. Complete the example.

$1\frac{3}{4}$ ◯ $1\frac{4}{5}$ ⇨ $\frac{}{4}$ ◯ $\frac{}{5}$ ⇨ $\frac{}{20}$ ◯ $\frac{}{20}$

C. Compare the mixed numbers.

$2\frac{1}{4}$ ◯ $3\frac{1}{3}$ $\qquad$ $5\frac{1}{2}$ ◯ $5\frac{3}{6}$ $\qquad$ $1\frac{2}{5}$ ◯ $1\frac{3}{5}$

$3\frac{2}{7}$ ◯ $3\frac{2}{9}$ $\qquad$ $1\frac{4}{6}$ ◯ $1\frac{6}{9}$ $\qquad$ $4\frac{7}{8}$ ◯ $6\frac{1}{8}$

$4\frac{5}{6}$ ◯ $4\frac{3}{4}$ $\qquad$ $6\frac{2}{4}$ ◯ $2\frac{5}{9}$ $\qquad$ $8\frac{6}{7}$ ◯ $8\frac{4}{5}$

Easy Peasy All-in-One Homeschool $\qquad$ EP Math 5/6 Workbook

LESSON 64 Practice

Compare the mixed numbers.

$2\dfrac{3}{8}\ \bigcirc\ 2\dfrac{1}{3}$ $\qquad$ $1\dfrac{3}{5}\ \bigcirc\ 2\dfrac{5}{9}$ $\qquad$ $3\dfrac{5}{7}\ \bigcirc\ 3\dfrac{4}{5}$

$5\dfrac{3}{4}\ \bigcirc\ 5\dfrac{5}{6}$ $\qquad$ $2\dfrac{4}{7}\ \bigcirc\ 2\dfrac{4}{8}$ $\qquad$ $1\dfrac{5}{9}\ \bigcirc\ 1\dfrac{5}{6}$

$7\dfrac{2}{5}\ \bigcirc\ 7\dfrac{2}{4}$ $\qquad$ $3\dfrac{1}{2}\ \bigcirc\ 5\dfrac{1}{3}$ $\qquad$ $4\dfrac{2}{3}\ \bigcirc\ 4\dfrac{4}{7}$

$2\dfrac{2}{3}\ \bigcirc\ 2\dfrac{4}{6}$ $\qquad$ $1\dfrac{1}{4}\ \bigcirc\ 1\dfrac{1}{9}$ $\qquad$ $7\dfrac{4}{5}\ \bigcirc\ 8\dfrac{1}{5}$

$6\dfrac{1}{7}\ \bigcirc\ 5\dfrac{3}{7}$ $\qquad$ $3\dfrac{6}{9}\ \bigcirc\ 3\dfrac{5}{8}$ $\qquad$ $2\dfrac{5}{6}\ \bigcirc\ 2\dfrac{4}{9}$

$7\dfrac{1}{2}\ \bigcirc\ 7\dfrac{3}{5}$ $\qquad$ $6\dfrac{1}{9}\ \bigcirc\ 6\dfrac{1}{3}$ $\qquad$ $5\dfrac{1}{2}\ \bigcirc\ 5\dfrac{4}{8}$

Easy Peasy All-in-One Homeschool $\qquad\qquad$ EP Math 5/6 Workbook

LESSON 65 Comparing Mixed and Improper Fractions

A. One way to compare a mixed number to an improper fraction is to convert the mixed number to an improper fraction and compare two improper fractions. Here is an example of using this method. Complete the example.

$$2\frac{1}{4} \bigcirc \frac{17}{6} \Rightarrow \frac{}{4} \bigcirc \frac{}{6} \Rightarrow \frac{}{12} \bigcirc \frac{}{12}$$

B. Another way to compare a mixed number to an improper fraction is to convert the improper fraction to a mixed number and compare two mixed numbers. This method helps you avoid dealing with large numbers. Complete the example below.

$$2\frac{1}{5} \bigcirc \frac{17}{8} \Rightarrow 2\frac{}{5} \bigcirc 2\frac{}{8} \Rightarrow \text{The whole number parts are the same, so } \frac{}{5} \bigcirc \frac{}{8}$$

C. Compare the fractions.

$$1\frac{3}{4} \bigcirc \frac{11}{8} \qquad \frac{19}{4} \bigcirc 3\frac{5}{6} \qquad 5\frac{2}{3} \bigcirc \frac{68}{12}$$

$$2\frac{2}{5} \bigcirc \frac{35}{11} \qquad \frac{15}{5} \bigcirc 3\frac{1}{8} \qquad 3\frac{5}{9} \bigcirc \frac{35}{12}$$

$$\frac{29}{13} \bigcirc 3\frac{1}{7} \qquad 2\frac{1}{5} \bigcirc \frac{44}{20} \qquad \frac{18}{10} \bigcirc 1\frac{2}{5}$$

$$4\frac{1}{2} \bigcirc \frac{27}{6} \qquad \frac{59}{9} \bigcirc 6\frac{7}{8} \qquad 3\frac{3}{4} \bigcirc \frac{26}{7}$$

Easy Peasy All-in-One Homeschool

EP Math 5/6 Workbook

LESSON 65 Practice

Compare the fractions.

$2\dfrac{4}{5}\ \bigcirc\ \dfrac{20}{3}$ $2\dfrac{1}{4}\ \bigcirc\ \dfrac{13}{6}$ $\dfrac{39}{12}\ \bigcirc\ 3\dfrac{1}{4}$

$\dfrac{24}{7}\ \bigcirc\ 2\dfrac{5}{6}$ $1\dfrac{8}{9}\ \bigcirc\ \dfrac{10}{3}$ $6\dfrac{3}{5}\ \bigcirc\ \dfrac{61}{12}$

$1\dfrac{9}{10}\ \bigcirc\ \dfrac{17}{9}$ $1\dfrac{3}{9}\ \bigcirc\ \dfrac{40}{30}$ $\dfrac{37}{9}\ \bigcirc\ 4\dfrac{1}{5}$

$\dfrac{75}{11}\ \bigcirc\ 9\dfrac{1}{7}$ $3\dfrac{4}{5}\ \bigcirc\ \dfrac{24}{9}$ $4\dfrac{3}{8}\ \bigcirc\ \dfrac{13}{3}$

$5\dfrac{8}{12}\ \bigcirc\ \dfrac{51}{9}$ $1\dfrac{5}{8}\ \bigcirc\ \dfrac{27}{15}$ $\dfrac{45}{7}\ \bigcirc\ 7\dfrac{3}{6}$

$\dfrac{85}{9}\ \bigcirc\ 6\dfrac{2}{5}$ $\dfrac{60}{7}\ \bigcirc\ 2\dfrac{11}{14}$ $\dfrac{96}{21}\ \bigcirc\ 4\dfrac{4}{7}$

Easy Peasy All-in-One Homeschool   EP Math 5/6 Workbook

LESSON 66 Adding Fractions with Like Denominators

A. To add fractions with like denominators, add the numerators and keep the same denominator. Then simplify the result and convert it to a mixed number, if possible. The following example shows this visually. Complete the example.

$$\frac{2}{9} + \frac{4}{9} = \frac{2+4}{9} = \frac{}{9} = \frac{}{3}$$

B. Add the fractions. Simplify and convert your answers to mixed numbers, if possible.

$$\frac{1}{4} + \frac{3}{4} =$$

$$\frac{2}{15} + \frac{10}{15} =$$

$$\frac{3}{8} + \frac{7}{8} =$$

$$\frac{8}{12} + \frac{6}{12} =$$

$$\frac{5}{6} + \frac{4}{6} =$$

$$\frac{17}{24} + \frac{9}{24} =$$

$$\frac{4}{9} + \frac{8}{9} =$$

$$\frac{15}{18} + \frac{12}{18} =$$

$$\frac{3}{5} + \frac{3}{5} =$$

$$\frac{11}{30} + \frac{14}{30} =$$

$$\frac{3}{28} + \frac{15}{28} + \frac{17}{28} =$$

$$\frac{34}{45} + \frac{42}{45} + \frac{20}{45} =$$

Easy Peasy All-in-One Homeschool   EP Math 5/6 Workbook

LESSON 66 Practice

Add the fractions. Simplify and convert your answers to mixed numbers, if possible.

$\dfrac{5}{6} + \dfrac{5}{6} =$ $\qquad$ $\dfrac{9}{16} + \dfrac{3}{16} =$

$\dfrac{5}{9} + \dfrac{7}{9} =$ $\qquad$ $\dfrac{5}{12} + \dfrac{11}{12} =$

$\dfrac{1}{4} + \dfrac{1}{4} =$ $\qquad$ $\dfrac{9}{11} + \dfrac{10}{11} =$

$\dfrac{6}{7} + \dfrac{3}{7} =$ $\qquad$ $\dfrac{12}{21} + \dfrac{6}{21} =$

$\dfrac{2}{3} + \dfrac{2}{3} =$ $\qquad$ $\dfrac{17}{60} + \dfrac{23}{60} =$

$\dfrac{7}{8} + \dfrac{5}{8} =$ $\qquad$ $\dfrac{15}{48} + \dfrac{37}{48} =$

$\dfrac{2}{5} + \dfrac{3}{5} =$ $\qquad$ $\dfrac{17}{32} + \dfrac{23}{32} =$

$\dfrac{8}{9} + \dfrac{7}{9} =$ $\qquad$ $\dfrac{24}{25} + \dfrac{11}{25} =$

Easy Peasy All-in-One Homeschool   EP Math 5/6 Workbook

LESSON 67 Subtracting Fractions with Like Denominators

A. To subtract fractions with like denominators, subtract the numerators and keep the same denominator. Then simplify the result, if possible. The following example shows this visually. Complete the example.

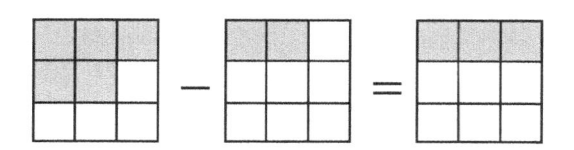

$$\frac{5}{9} - \frac{2}{9} = \frac{5-2}{9} = \frac{}{9} = \frac{}{3}$$

B. Subtract the fractions. Simplify your answers to the lowest terms.

$$\frac{2}{6} - \frac{0}{6} =$$

$$\frac{9}{10} - \frac{1}{10} =$$

$$\frac{4}{5} - \frac{2}{5} =$$

$$\frac{14}{18} - \frac{8}{18} =$$

$$\frac{6}{7} - \frac{4}{7} =$$

$$\frac{14}{15} - \frac{5}{15} =$$

$$\frac{4}{9} - \frac{1}{9} =$$

$$\frac{11}{12} - \frac{7}{12} =$$

$$\frac{1}{2} - \frac{1}{2} =$$

$$\frac{19}{24} - \frac{9}{24} =$$

$$\frac{7}{8} - \frac{3}{8} =$$

$$\frac{28}{45} - \frac{16}{45} =$$

$$\frac{5}{6} - \frac{2}{6} =$$

$$\frac{27}{36} - \frac{15}{36} =$$

Easy Peasy All-in-One Homeschool

EP Math 5/6 Workbook

LESSON 67 Practice

Subtract the fractions. Simplify your answers to the lowest terms.

$\dfrac{8}{9} - \dfrac{3}{9} =$ $\qquad$ $\dfrac{10}{12} - \dfrac{7}{12} =$

$\dfrac{4}{6} - \dfrac{1}{6} =$ $\qquad$ $\dfrac{11}{15} - \dfrac{2}{15} =$

$\dfrac{7}{8} - \dfrac{5}{8} =$ $\qquad$ $\dfrac{17}{18} - \dfrac{8}{18} =$

$\dfrac{3}{5} - \dfrac{2}{5} =$ $\qquad$ $\dfrac{13}{14} - \dfrac{6}{14} =$

$\dfrac{5}{7} - \dfrac{0}{7} =$ $\qquad$ $\dfrac{14}{16} - \dfrac{8}{16} =$

$\dfrac{3}{4} - \dfrac{1}{4} =$ $\qquad$ $\dfrac{17}{20} - \dfrac{5}{20} =$

$\dfrac{5}{9} - \dfrac{2}{9} =$ $\qquad$ $\dfrac{29}{33} - \dfrac{14}{33} =$

$\dfrac{4}{8} - \dfrac{4}{8} =$ $\qquad$ $\dfrac{67}{84} - \dfrac{46}{84} =$

Easy Peasy All-in-One Homeschool $\qquad$ EP Math 5/6 Workbook

LESSON 68 Finding the Least Common Multiple

A. A **multiple** is a number you get when you multiply by an integer. For example, the multiples of 3 are 3, 6, 9, 12, and so on. Circle the numbers below that are multiples of 6.

| 12 | 24 | 46 | 18 | 30 | 52 | 16 | 48 | 32 | 20 |

B. The **least common multiple (LCM)** of two integers is the smallest positive integer that is a multiple of both integers. We have already learned two methods for finding the LCM: the listing method and the prime factorization method. Find the LCM of 9 and 12 using one of the two methods. Show your steps. Review Lesson 49 if needed.

C. Find the least common multiple of each number pair.

6   4   3
8   9   7

5   8   9
10   12   15

16   12   15
28   30   25

24   36   64
28   45   96

Easy Peasy All-in-One Homeschool   EP Math 5/6 Workbook

LESSON 68 Practice

Find the least common multiple of each number pair.

2
3

6
9

7
8

4
10

9
12

6
11

12
15

13
39

18
21

10
55

14
42

11
30

27
42

25
35

32
40

49
56

36
54

48
88

Easy Peasy All-in-One Homeschool   EP Math 5/6 Workbook

LESSON 69 Finding Common Denominators

A. When two or more fractions have the same denominator, we say they have **common denominators**. Fractions with common denominators are easy to add, subtract, and compare because we can simply add, subtract, and compare their numerators. That is why, when we deal with fractions with different denominators, we rewrite them as equivalent fractions with common denominators.

B. When rewriting fractions, you can use any of their common denominators. Use the easiest one to work with. Here is an example of rewriting fractions with three different common denominators. Complete the example.

$$\frac{4}{5} \ \& \ \frac{7}{10} \iff \frac{}{10} \ \& \ \frac{}{10} \iff \frac{}{20} \ \& \ \frac{}{20} \iff \frac{}{50} \ \& \ \frac{}{50}$$

C. The **least common denominator (LCD)** is often used as a common denominator because it is easier to work with smaller numbers. The least common denominator is the least common multiple (LCM) of the denominators. Here is an example of rewriting fractions with the least common denominator. Complete the example.

$$\frac{2}{6} \ \& \ \frac{4}{15} \implies$$ Find the LCD or LCM of 6 and 15 (see Lesson 49). A quick tip is to check multiples of 15 until you find one that is also a multiple of 6. 30 is the one! $$\implies \frac{}{30} \ \& \ \frac{}{30}$$

D. Rewrite each pair of fractions with the least common denominator.

$$\frac{7}{9}, \frac{9}{12} \qquad\qquad\qquad \frac{7}{15}, \frac{3}{10}$$

$$\frac{3}{8}, \frac{5}{20} \qquad\qquad\qquad \frac{5}{16}, \frac{9}{24}$$

$$\frac{5}{6}, \frac{10}{21} \qquad\qquad\qquad \frac{11}{12}, \frac{13}{30}$$

Easy Peasy All-in-One Homeschool — EP Math 5/6 Workbook

LESSON 69 Practice

A. Rewrite each pair of fractions with two different common denominators.

$\dfrac{1}{2}, \dfrac{4}{5}$ _____ _____

$\dfrac{2}{3}, \dfrac{4}{15}$ _____ _____

$\dfrac{3}{4}, \dfrac{5}{6}$ _____ _____

$\dfrac{3}{4}, \dfrac{7}{10}$ _____ _____

$\dfrac{2}{5}, \dfrac{3}{7}$ _____ _____

$\dfrac{1}{9}, \dfrac{5}{12}$ _____ _____

B. Rewrite each pair of fractions with the least common denominator.

$\dfrac{2}{3}, \dfrac{6}{7}$

$\dfrac{7}{12}, \dfrac{11}{18}$

$\dfrac{1}{4}, \dfrac{5}{18}$

$\dfrac{9}{14}, \dfrac{13}{35}$

$\dfrac{7}{6}, \dfrac{9}{15}$

$\dfrac{16}{27}, \dfrac{37}{54}$

$\dfrac{4}{9}, \dfrac{5}{16}$

$\dfrac{20}{21}, \dfrac{15}{28}$

Easy Peasy All-in-One Homeschool — EP Math 5/6 Workbook

LESSON 70 Adding Fractions with Unlike Denominators

A. To add fractions with unlike denominators, rewrite the fractions as equivalent fractions with common denominators and add their numerators. Then simplify the result and convert it to a mixed number, if possible. Here is an example of rewriting fractions with the least common denominator (LCD) before adding. Complete the example.

$$\frac{5}{9} + \frac{7}{15} \implies \text{The LCD or LCM of 9 and 15 is 45 (see Lessons 49 and 69).} \implies \frac{25}{45} + \frac{}{45} = \frac{}{45} = 1\frac{}{45}$$

B. Simplifying fractions before adding can make the process easier since you'll be dealing with smaller numbers. See if you can reduce! Complete the example below.

$$\frac{6}{9} + \frac{12}{15} = \frac{2}{3} + \frac{}{5} = \frac{10}{15} + \frac{}{15} = \frac{}{15} = 1\frac{}{15}$$

C. Add the fractions. Simplify and convert your answers to mixed numbers, if possible.

$$\frac{2}{3} + \frac{5}{9} =$$

$$\frac{20}{24} + \frac{15}{18} =$$

$$\frac{4}{8} + \frac{6}{7} =$$

$$\frac{18}{27} + \frac{10}{12} =$$

$$\frac{6}{9} + \frac{3}{4} =$$

$$\frac{12}{20} + \frac{11}{15} =$$

$$\frac{1}{2} + \frac{7}{8} =$$

$$\frac{21}{24} + \frac{14}{16} =$$

Easy Peasy All-in-One Homeschool — EP Math 5/6 Workbook

LESSON 70 Practice

Add the fractions. Simplify and convert your answers to mixed numbers, if possible.

$\dfrac{1}{3} + \dfrac{1}{2} =$ $\qquad\qquad\qquad$ $\dfrac{2}{4} + \dfrac{9}{20} =$

$\dfrac{5}{8} + \dfrac{3}{4} =$ $\qquad\qquad\qquad$ $\dfrac{4}{5} + \dfrac{8}{12} =$

$\dfrac{4}{6} + \dfrac{1}{4} =$ $\qquad\qquad\qquad$ $\dfrac{2}{7} + \dfrac{10}{25} =$

$\dfrac{8}{9} + \dfrac{2}{6} =$ $\qquad\qquad\qquad$ $\dfrac{4}{18} + \dfrac{22}{36} =$

$\dfrac{3}{9} + \dfrac{4}{7} =$ $\qquad\qquad\qquad$ $\dfrac{20}{45} + \dfrac{15}{30} =$

$\dfrac{4}{6} + \dfrac{5}{8} =$ $\qquad\qquad\qquad$ $\dfrac{12}{21} + \dfrac{14}{28} =$

Easy Peasy All-in-One Homeschool $\qquad\qquad\qquad$ EP Math 5/6 Workbook

LESSON 71 Subtracting Fractions with Unlike Denominators

A. To subtract fractions with unlike denominators, rewrite the fractions as equivalent fractions with common denominators and subtract their numerators. Then simplify the result, if possible. Here is an example of rewriting fractions with the least common denominator (LCD) before subtracting. Complete the example.

$$\frac{7}{10} - \frac{7}{14} \quad \Rightarrow \quad \text{The LCD or LCM of 10 and 14 is 70 (see Lessons 49 and 69).} \quad \Rightarrow \quad \frac{49}{70} - \frac{}{70} = \frac{}{70} = \frac{}{5}$$

B. Simplifying fractions before subtracting can make the process easier since you'll be dealing with smaller numbers. See if you can reduce! Complete the example below.

$$\frac{8}{10} - \frac{6}{14} = \frac{4}{5} - \frac{}{7} = \frac{28}{35} - \frac{}{35} = \frac{}{35}$$

C. Subtract the fractions. Simplify your answers to the lowest terms.

$$\frac{2}{3} - \frac{1}{5} = \qquad\qquad \frac{11}{15} - \frac{8}{12} =$$

$$\frac{4}{7} - \frac{1}{2} = \qquad\qquad \frac{9}{16} - \frac{8}{24} =$$

$$\frac{6}{8} - \frac{2}{4} = \qquad\qquad \frac{24}{28} - \frac{16}{48} =$$

$$\frac{5}{6} - \frac{2}{5} = \qquad\qquad \frac{16}{20} - \frac{14}{35} =$$

Easy Peasy All-in-One Homeschool   EP Math 5/6 Workbook

LESSON 71 Practice

Subtract the fractions. Simplify your answers to the lowest terms.

$\dfrac{1}{2} - \dfrac{1}{3} =$

$\dfrac{9}{10} - \dfrac{5}{15} =$

$\dfrac{4}{6} - \dfrac{3}{9} =$

$\dfrac{7}{12} - \dfrac{3}{16} =$

$\dfrac{4}{5} - \dfrac{1}{2} =$

$\dfrac{17}{18} - \dfrac{15}{45} =$

$\dfrac{5}{8} - \dfrac{1}{3} =$

$\dfrac{16}{20} - \dfrac{18}{24} =$

$\dfrac{3}{4} - \dfrac{1}{5} =$

$\dfrac{32}{49} - \dfrac{20}{35} =$

$\dfrac{5}{9} - \dfrac{2}{4} =$

$\dfrac{54}{60} - \dfrac{40}{50} =$

Easy Peasy All-in-One Homeschool

EP Math 5/6 Workbook

LESSON 72 Word Problems: Adding/Subtracting Fractions

Solve. Make sure you check what question is being asked before you finish.

Yesterday Emily ran one half of a 5-mile trail. Today she ran two fifths of the trail. On which day did she run farther? _____

Sarah picked three and two-thirds baskets of apples. Chris picked four and a quarter baskets of apples. Who picked more apples? _____

On Monday Percy read three eighths of his reading assignment. On Tuesday he read the rest. On which day did he read more? _____

Alan and Betty had a pizza for dinner. Alan ate $1/3$ of the pizza. Betty ate $2/5$ of the pizza. What fraction of the pizza was eaten? _____

There was $7/9$ of a bottle of milk. Danny drank some and $1/3$ of the bottle was left. What fraction of the bottle did Danny drink? _____

Carol read $2/6$ of a book yesterday and $5/12$ of the book today. What fraction of the book did she read in all? _____

Sarah spent $2/5$ of her allowance on books and $3/8$ on snacks. What fraction of the allowance did she have <u>left</u>? _____

Holly finished $3/4$ of the homework. Evan finished $8/9$ of the homework. How much more of the homework did Evan finish? _____

Noah had a box of chocolates. He ate $2/8$ of the box and gave $2/3$ to his sister. What fraction of the box was left over? _____

Emily had $5/6$ of a carton of eggs. She used $1/3$ of the carton to bake cookies. What fraction of the carton was left over? _____

Jessica painted $7/10$ of her room. Her brother painted $1/5$ of the room. What fraction of the room was not painted yet? _____

A recipe calls for $3/4$ cup of flour to bake a pie. Dana has only $1/2$ cup of flour. How much more flour does Dana need? _____

Easy Peasy All-in-One Homeschool   EP Math 5/6 Workbook

LESSON 72 Practice

Solve. Make sure you check what question is being asked before you finish.

Laura walked two and three-fifths miles. Justin walked two and one-half miles. Who walked the farther distance? _____

Lincoln solved three fourths of the problems on the worksheet. Angie solved five sevenths of the problems. Who solved more? _____

Noah sleeps seven and one-half hours on average. Matt sleeps eight hours on average. Who sleeps more hours on average? _____

In a library, $2/5$ of the books are fiction and the rest are non-fiction. What fraction of the books are non-fiction? _____

Matt drank $1/6$ of a bottle of milk. Now the bottle is $1/4$ full. What fraction of the whole bottle was there at first? _____

Jasmine used $2/3$ cup of flour and $1/2$ cup of sugar to bake cookies. How much more flour than sugar was used? _____

Jack and Sam had a pizza for dinner. Jack ate $2/5$ of the pizza. Sam ate $3/10$ of the pizza. What fraction of the pizza was left? _____

A farmer planted corn on $4/10$ of his field and beans on $4/12$ of the field. What fraction of the field was planted in total? _____

Olivia painted $1/5$ of her room yellow and $1/4$ of the room ivory. What fraction of the room was not painted yet? _____

Kyle filled $5/9$ of a bucket with water. Amy filled $2/6$ of the bucket. What fraction of the bucket was filled with water? _____

Tylor read $4/12$ of a book yesterday. He read $3/8$ of the book today. How much more of the book did he read today? _____

Maria spent $1/3$ of her money on a book and $2/9$ on a video game. What fraction of her money did she spend in total? _____

Easy Peasy All-in-One Homeschool   EP Math 5/6 Workbook

LESSON 73 Adding Negative Fractions

A. Use the examples below to explain how to add integers. These examples are from Lesson 16. Review that lesson if you need help.

$-4 + 7 =$ $\qquad$ $4 + -7 =$ $\qquad$ $-4 + -7 =$

B. Just like integers, fractions can be negative. A negative sign can be in the numerator, in the denominator, or in front of the entire fraction. The following examples show three ways to write a negative fraction. Complete the second example.

$$-\frac{3}{4} = \frac{-3}{4} = \frac{3}{-4} \qquad\qquad -\frac{11}{15} = \frac{}{15} = \frac{11}{}$$

C. You can add negative fractions just as you add integers. Complete the second example.

$$\frac{3}{4} + \left(-\frac{5}{6}\right) = \frac{3}{4} - \frac{5}{6} = \frac{9}{12} - \frac{10}{12} = -\frac{1}{12}$$

$$-\frac{4}{5} + \left(-\frac{2}{3}\right) = -\left(\frac{4}{5} + \frac{2}{3}\right) = -\left(\frac{}{15} + \frac{}{15}\right) = -\frac{}{15} = -1\frac{}{15}$$

D. Add the fractions. Simplify and convert your answers to mixed numbers.

$-\dfrac{2}{9} + \dfrac{4}{9} =$ $\qquad\qquad$ $-\dfrac{5}{7} + \left(-\dfrac{6}{7}\right) =$

$-\dfrac{2}{5} + \dfrac{1}{4} =$ $\qquad\qquad$ $\dfrac{2}{3} + \left(-\dfrac{5}{7}\right) =$

$\dfrac{3}{4} + \left(-\dfrac{1}{3}\right) =$ $\qquad\qquad$ $-\dfrac{3}{8} + \left(-\dfrac{5}{6}\right) =$

Easy Peasy All-in-One Homeschool $\qquad\qquad$ EP Math 5/6 Workbook

LESSON 73 Practice

Add the fractions. Simplify and convert your answers to mixed numbers.

$-\dfrac{3}{8} + \dfrac{5}{8} =$ $-\dfrac{5}{6} + \left(-\dfrac{1}{6}\right) =$

$-\dfrac{5}{8} + \dfrac{1}{4} =$ $\dfrac{2}{3} + \left(-\dfrac{3}{4}\right) =$

$-\dfrac{2}{5} + \dfrac{1}{4} =$ $-\dfrac{1}{3} + \left(-\dfrac{1}{9}\right) =$

$-\dfrac{3}{4} + \dfrac{2}{6} =$ $\dfrac{7}{8} + \left(-\dfrac{1}{2}\right) =$

$-\dfrac{2}{3} + \left(-\dfrac{1}{6}\right) =$ $\dfrac{3}{5} + \left(-\dfrac{1}{2}\right) =$

$\dfrac{2}{7} + \left(-\dfrac{5}{6}\right) =$ $-\dfrac{5}{9} + \left(-\dfrac{2}{3}\right) =$

Easy Peasy All-in-One Homeschool   EP Math 5/6 Workbook

LESSON 74 Subtracting Negative Fractions

A. Use the examples below to explain how to subtract integers. These examples are from Lesson 14. Review that lesson if you need help.

$3 - -7 =$ $\qquad$ $-9 - 2 =$ $\qquad$ $-8 - -5 =$

B. You can subtract negative fractions just as you subtract integers. Remember that subtraction is the opposite of addition. Complete the examples.

$$\frac{2}{3} - \left(-\frac{1}{2}\right) = \frac{2}{3} + \frac{1}{2} = \frac{}{6} + \frac{}{6} = \frac{}{6} = 1\frac{}{6}$$

$$-\frac{3}{8} - \left(-\frac{1}{6}\right) = -\frac{3}{8} + \frac{1}{6} = \frac{1}{6} - \frac{3}{8} = \frac{}{24} - \frac{}{24} = -\frac{}{24}$$

$$-\frac{4}{15} - \frac{1}{12} = -\left(\frac{4}{15} + \frac{1}{12}\right) = -\left(\frac{}{60} + \frac{}{60}\right) = -\frac{}{60} = -\frac{}{20}$$

C. Subtract the fractions. Simplify and convert your answers to mixed numbers.

$-\dfrac{3}{5} - \dfrac{2}{5} =$ $\qquad\qquad$ $-\dfrac{2}{9} - \left(-\dfrac{4}{9}\right) =$

$-\dfrac{1}{2} - \dfrac{3}{8} =$ $\qquad\qquad$ $\dfrac{1}{5} - \left(-\dfrac{5}{6}\right) =$

$\dfrac{3}{4} - \left(-\dfrac{1}{2}\right) =$ $\qquad\qquad$ $-\dfrac{7}{9} - \left(-\dfrac{5}{12}\right) =$

Easy Peasy All-in-One Homeschool $\qquad$ EP Math 5/6 Workbook

LESSON 74 Practice

Subtract the fractions. Simplify and convert your answers to mixed numbers.

$-\dfrac{3}{4} - \dfrac{3}{4} =$ $\qquad$ $\dfrac{4}{9} - \left(-\dfrac{8}{9}\right) =$

$\dfrac{1}{3} - \dfrac{3}{4} =$ $\qquad$ $-\dfrac{2}{5} - \left(-\dfrac{4}{6}\right) =$

$-\dfrac{4}{7} - \dfrac{5}{6} =$ $\qquad$ $\dfrac{8}{9} - \left(-\dfrac{5}{6}\right) =$

$-\dfrac{1}{6} + \dfrac{3}{4} =$ $\qquad$ $-\dfrac{2}{5} - \left(-\dfrac{1}{3}\right) =$

$-\dfrac{5}{12} - \dfrac{7}{8} =$ $\qquad$ $-\dfrac{1}{6} - \left(-\dfrac{11}{18}\right) =$

$\dfrac{3}{10} - \dfrac{8}{15} =$ $\qquad$ $\dfrac{7}{12} - \left(-\dfrac{5}{15}\right) =$

Easy Peasy All-in-One Homeschool $\qquad$ EP Math 5/6 Workbook

LESSON 75 Fractions on a Number Line

A. On a number line, proper fractions are located between 0 and 1. Here are some proper fractions placed on number lines. Fill in the missing fractions.

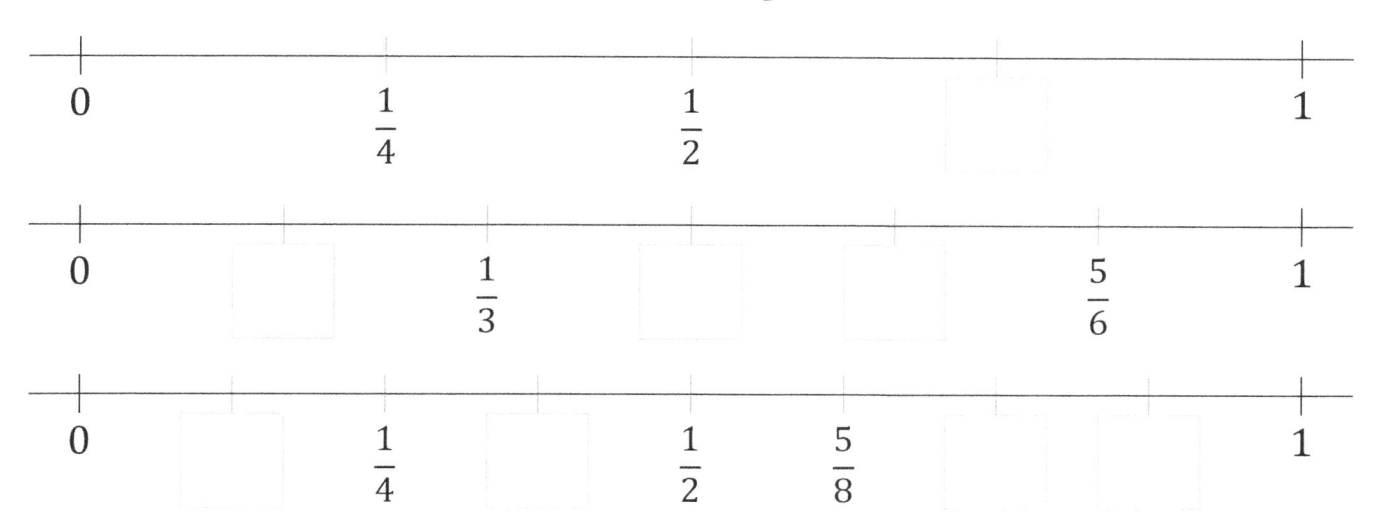

```
0 ——— 1/4 ——— 1/2 ——— ☐ ——— 1
```

```
0 ——— ☐ ——— 1/3 ——— ☐ ——— ☐ ——— 5/6 ——— 1
```

```
0 ——— ☐ ——— 1/4 ——— ☐ ——— 1/2 ——— 5/8 ——— ☐ ——— ☐ ——— 1
```

B. Match the location of each letter with the correct fraction. Count how many parts it is divided into. That is the denominator, but the fractions below have been reduced.

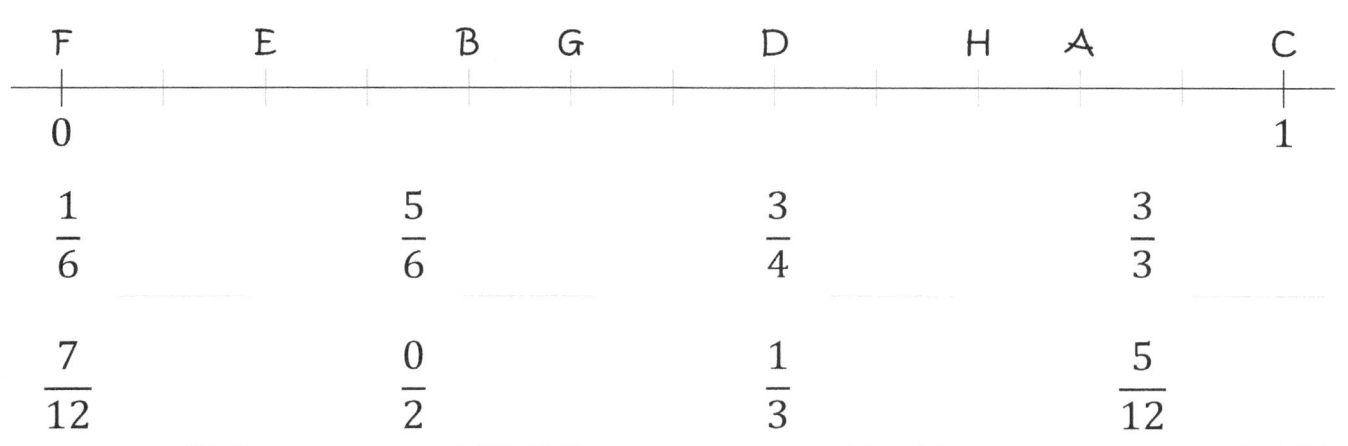

```
F   E   B  G     D     H  A        C
0                                  1
```

$\dfrac{1}{6}$ $\dfrac{5}{6}$ $\dfrac{3}{4}$ $\dfrac{3}{3}$

$\dfrac{7}{12}$ $\dfrac{0}{2}$ $\dfrac{1}{3}$ $\dfrac{5}{12}$

C. Write the fractions marked on the number line. Simplify your answers.

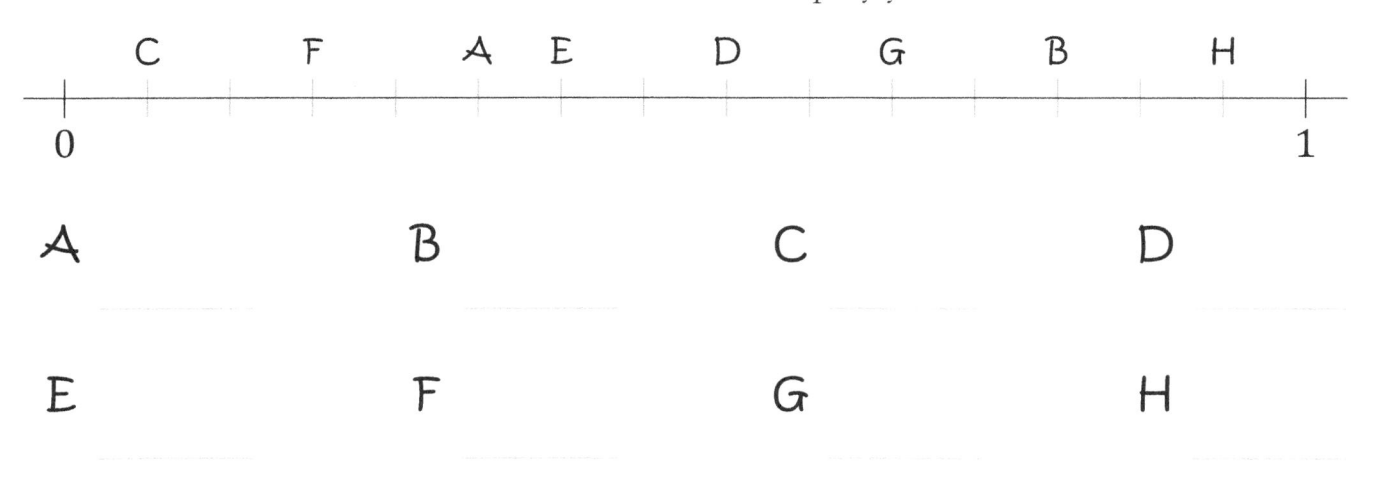

```
    C   F   A E   D   G   B   H
0                               1
```

A   B   C   D

E   F   G   H

Easy Peasy All-in-One Homeschool   EP Math 5/6 Workbook

LESSON 75 Practice

A. Place each set of fractions on the number line in the correct order.

$\dfrac{3}{4}$ $\dfrac{7}{8}$ $\dfrac{1}{2}$ $\dfrac{1}{8}$

$\dfrac{3}{5}$ $\dfrac{1}{10}$ $\dfrac{1}{5}$ $\dfrac{9}{10}$

$\dfrac{1}{3}$ $\dfrac{4}{9}$ $\dfrac{2}{3}$ $\dfrac{1}{9}$

$\dfrac{7}{10}$ $\dfrac{4}{5}$ $\dfrac{3}{20}$ $\dfrac{3}{10}$

$\dfrac{2}{9}$ $\dfrac{2}{3}$ $\dfrac{5}{6}$ $\dfrac{1}{2}$

$\dfrac{5}{6}$ $\dfrac{2}{3}$ $\dfrac{3}{8}$ $\dfrac{1}{12}$

B. Write the fractions represented on the number line. Simplify your answers.

A.

B.

C.

D.

E.

F.

G.

H.

Easy Peasy All-in-One Homeschool   EP Math 5/6 Workbook

LESSON 76 Fractions on a Number Line

A. Improper and mixed fractions are located between integers on a number line, as shown in the examples below. Fill in the missing fractions.

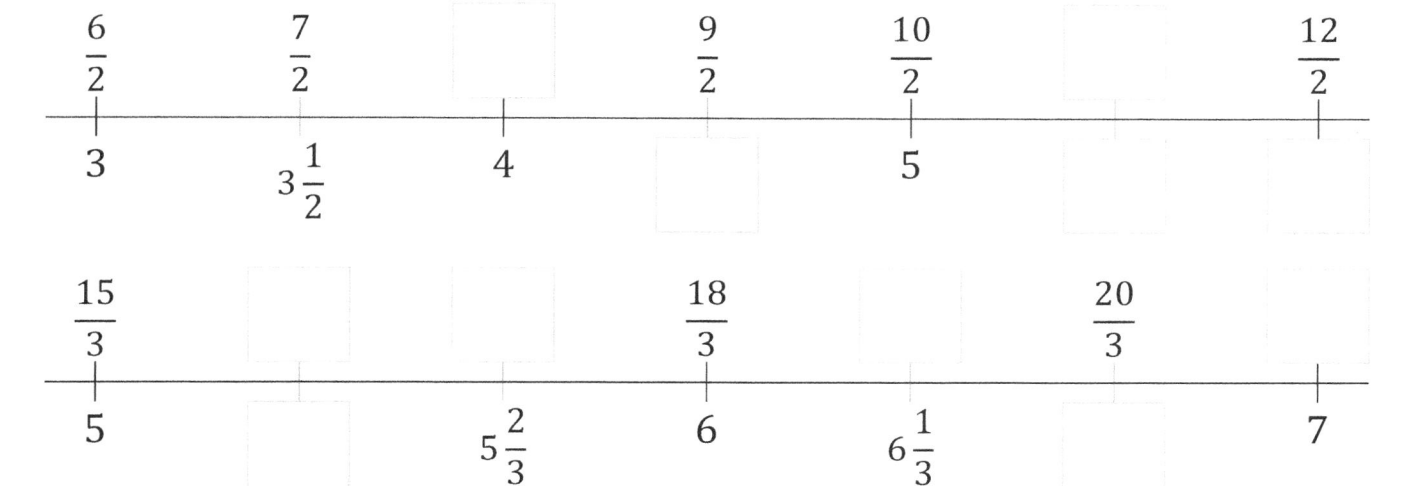

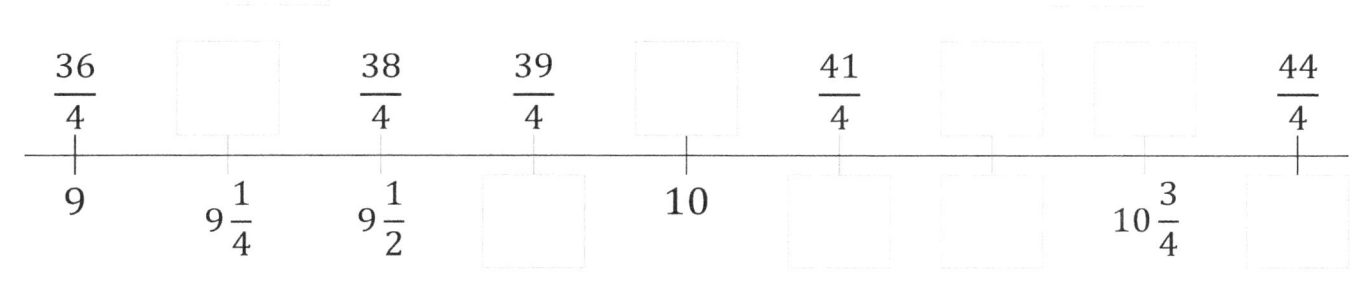

$\frac{6}{2}$ $\frac{7}{2}$ ☐ $\frac{9}{2}$ $\frac{10}{2}$ ☐ $\frac{12}{2}$

3    $3\frac{1}{2}$ 4   ☐   5   ☐   ☐

$\frac{15}{3}$ ☐   ☐ $\frac{18}{3}$ ☐ $\frac{20}{3}$ ☐

5   ☐ $5\frac{2}{3}$ 6 $6\frac{1}{3}$ ☐   7

$\frac{36}{4}$ ☐ $\frac{38}{4}$ $\frac{39}{4}$ ☐ $\frac{41}{4}$ ☐ $\frac{44}{4}$

9    $9\frac{1}{4}$ $9\frac{1}{2}$ ☐   10   ☐ $10\frac{3}{4}$ ☐

B. Write each fraction marked on the number line as both a mixed fraction and an improper fraction. Reduce your answers to their lowest terms. It is divided into sixths.

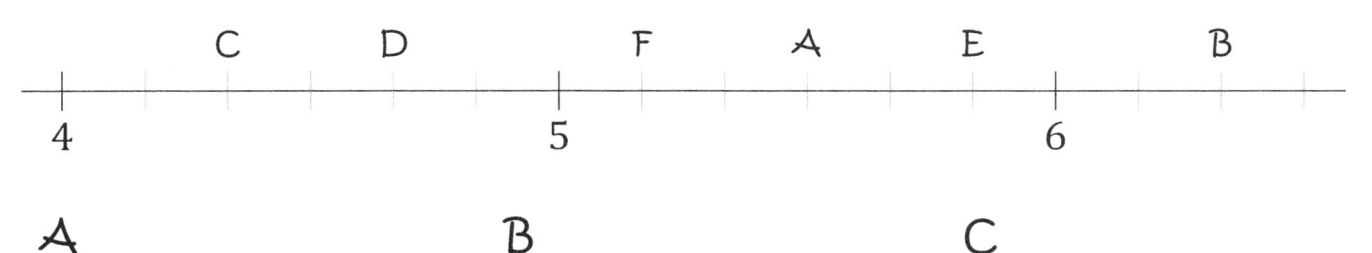

Number line from 4 to 6 with points: C, D, F, A, E, B

A.

B.

C.

D.

E.

F.

A math riddle for you! I am making treat bags. If I put in one apple per bag, I have one apple too many. If I put in two apples per bag, I have one bag too many. How many treat bags and apples do I have?

Easy Peasy All-in-One Homeschool   EP Math 5/6 Workbook

LESSON 76 Practice

A. Write each fraction marked on the number line as both a mixed fraction and an improper fraction. Reduce your answers to their lowest terms.

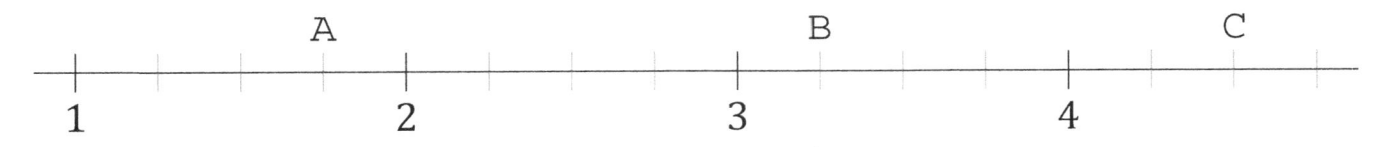

A

B

C

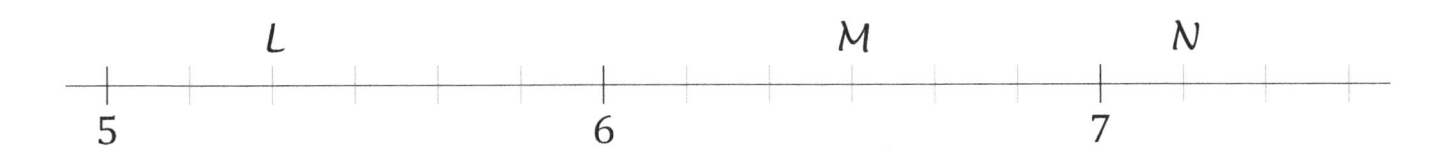

L

M

N

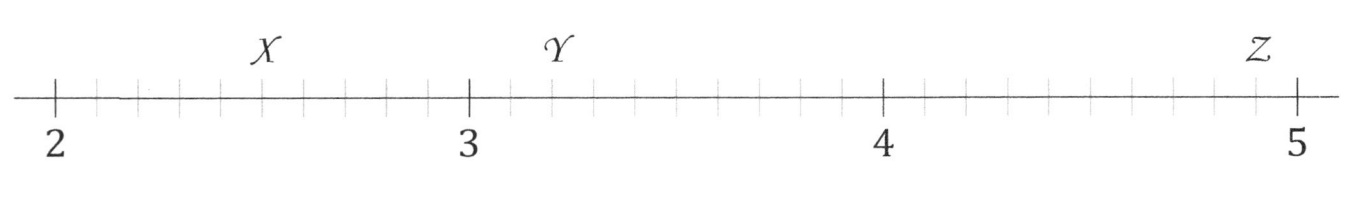

X

Y

Z

B. Convert between improper fractions and mixed numbers.

$2\dfrac{4}{7} =$ $5\dfrac{1}{6} =$ $\dfrac{38}{5} =$

$3\dfrac{1}{5} =$ $6\dfrac{5}{8} =$ $\dfrac{40}{7} =$

$8\dfrac{2}{3} =$ $4\dfrac{7}{9} =$ $\dfrac{79}{8} =$

Easy Peasy All-in-One Homeschool   EP Math 5/6 Workbook

LESSON 77 Negative Fractions on a Number Line

A. Positive fractions are located to the right of zero and negative fractions are located to the left of zero on a number line. Here are some fractions with different signs placed on number lines. Fill in the missing fractions.

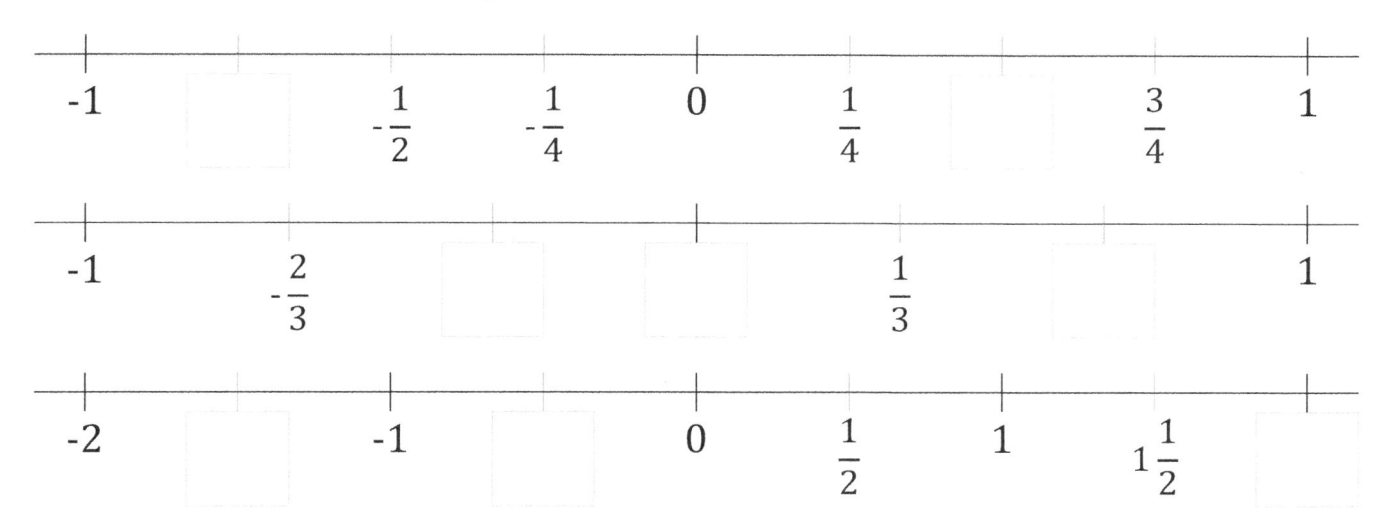

Number line 1: -1, [ ], -1/2, -1/4, 0, 1/4, [ ], 3/4, 1

Number line 2: -1, -2/3, [ ], [ ], 1/3, [ ], 1

Number line 3: -2, [ ], -1, [ ], 0, 1/2, 1, 1 1/2, [ ]

B. Match the location of each letter with the correct fraction. How many parts between each number?

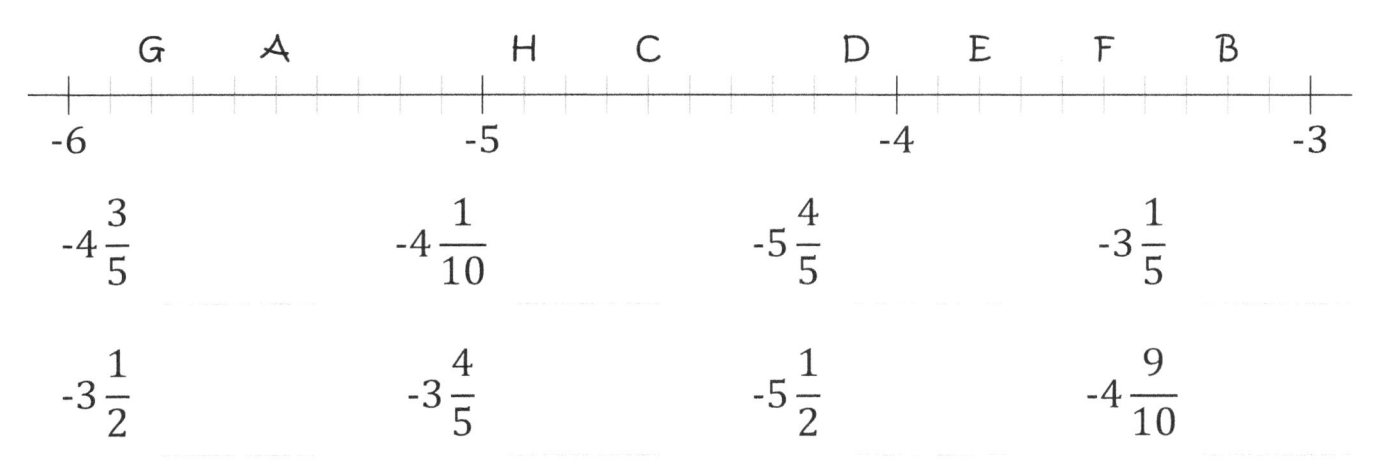

Number line with letters G, A, H, C, D, E, F, B from -6 to -3.

$-4\frac{3}{5}$ $-4\frac{1}{10}$ $-5\frac{4}{5}$ $-3\frac{1}{5}$

$-3\frac{1}{2}$ $-3\frac{4}{5}$ $-5\frac{1}{2}$ $-4\frac{9}{10}$

C. Use the number line below to order each set of fractions from least to greatest.

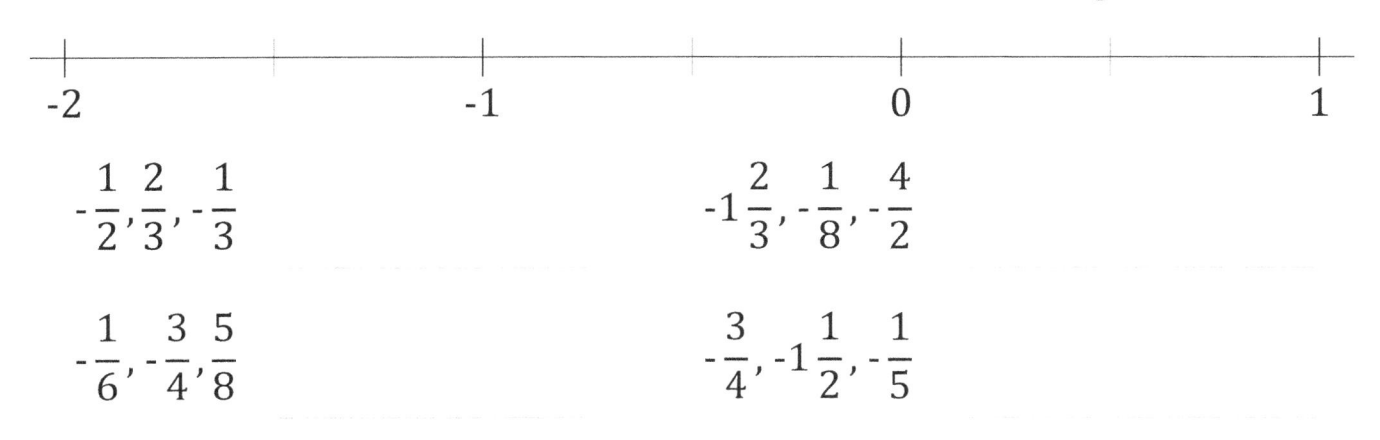

Number line from -2 to 1.

$-\frac{1}{2}, \frac{2}{3}, -\frac{1}{3}$ $-1\frac{2}{3}, -\frac{1}{8}, -\frac{4}{2}$

$-\frac{1}{6}, -\frac{3}{4}, \frac{5}{8}$ $-\frac{3}{4}, -1\frac{1}{2}, -\frac{1}{5}$

Easy Peasy All-in-One Homeschool   EP Math 5/6 Workbook

LESSON 77 Practice

A. Place each set of fractions on the number line in the correct order.

$\dfrac{3}{4}$ $-\dfrac{1}{4}$ $-\dfrac{3}{4}$ ⊢—————|—————|—————⊢ -1   0   1

$-\dfrac{6}{5}$ $-\dfrac{9}{5}$ $-\dfrac{3}{5}$ ⊢—————|—————|—————⊢ -2   -1   0

$-4\dfrac{2}{3}$ $-\dfrac{7}{3}$ $-\dfrac{10}{3}$ ⊢———|———|———|———⊢ -5   -4   -3   -2

$-\dfrac{5}{2}$ $-\dfrac{10}{8}$ $-\dfrac{11}{4}$ ⊢—————|—————|—————⊢ -3   -2   -1

$-\dfrac{4}{6}$ $-\dfrac{5}{2}$ $\dfrac{9}{18}$ ⊢——|——|——|——|——⊢ -3   -2   -1   0   1

$-7\dfrac{2}{9}$ $-8\dfrac{5}{9}$ $-\dfrac{23}{3}$ ⊢—————|—————|—————⊢ -9   -8   -7

B. Match the location of each letter with the correct fraction.

```
      A  H         F       E      B       D      G       C
⊢——|——|——|——|——|——|——|——|——|——|——|——|——|——|——|——|——|——⊢
-4                 -3                    -2
```

$-1\dfrac{5}{6}$ $-2\dfrac{5}{6}$ $-\dfrac{23}{6}$ $-3\dfrac{1}{6}$

$-\dfrac{3}{2}$ $-\dfrac{11}{3}$ $-2\dfrac{1}{2}$ $-\dfrac{13}{6}$

Easy Peasy All-in-One Homeschool   EP Math 5/6 Workbook

LESSON 78 Adding Mixed Numbers

A. To add mixed numbers, add the whole number parts and the fraction parts separately and then combine the two parts. Study the example carefully!

$$1\frac{4}{7} + 2\frac{1}{7} = (1+2) + \left(\frac{4}{7} + \frac{1}{7}\right) = 3 + \frac{5}{7} = 3\frac{5}{7}$$

B. Always write your answer in simplest form. Simplify the sum of the fraction parts (if possible) and convert it to a mixed number (if improper) before combining.

$$2\frac{3}{8} + 3\frac{7}{8} = 5 + \frac{10}{8} = 5 + \frac{5}{4} = 5 + 1\frac{1}{4} = 6\frac{1}{4}$$

C. To add mixed numbers with unlike denominators, rewrite them as equivalent fractions with common denominators and then add the whole number parts and the fraction parts separately. Be sure to write your answer in simplest form. Complete the example.

$$2\frac{3}{4} + 3\frac{4}{5} = 2\frac{15}{20} + 3\frac{16}{20} = 5 + \frac{}{20} = 5 + 1\frac{}{20} = 6\frac{}{20}$$

D. Add the mixed numbers. Simplify and convert your answers to mixed numbers.

$$2\frac{1}{5} + 1\frac{2}{5} = \qquad\qquad 3\frac{7}{9} + 4\frac{8}{9} =$$

$$4\frac{1}{2} + 2\frac{2}{5} = \qquad\qquad 4\frac{2}{3} + 3\frac{1}{2} =$$

$$7\frac{5}{6} + 1\frac{1}{3} = \qquad\qquad 2\frac{5}{8} + 5\frac{3}{4} =$$

Easy Peasy All-in-One Homeschool — EP Math 5/6 Workbook

LESSON 78 Practice

Add the mixed numbers. Simplify and convert your answers to mixed numbers.

$3\dfrac{1}{9} + 4\dfrac{5}{9} =$ $\qquad$ $1\dfrac{3}{8} + 5\dfrac{3}{8} =$

$2\dfrac{4}{5} + 6\dfrac{2}{5} =$ $\qquad$ $4\dfrac{5}{7} + 3\dfrac{4}{7} =$

$2\dfrac{1}{4} + 7\dfrac{1}{2} =$ $\qquad$ $1\dfrac{1}{6} + 3\dfrac{2}{3} =$

$4\dfrac{1}{2} + 2\dfrac{3}{5} =$ $\qquad$ $2\dfrac{3}{4} + 2\dfrac{2}{3} =$

$5\dfrac{2}{3} + 3\dfrac{5}{9} =$ $\qquad$ $1\dfrac{5}{6} + 6\dfrac{3}{4} =$

$4\dfrac{5}{6} + 2\dfrac{3}{8} =$ $\qquad$ $3\dfrac{4}{5} + 3\dfrac{2}{7} =$

Easy Peasy All-in-One Homeschool   EP Math 5/6 Workbook

LESSON 79 Subtracting Mixed Numbers

A. To subtract mixed numbers, subtract the whole number parts and the fraction parts separately and then combine the two parts. Complete the example.

$$5\frac{7}{9} - 3\frac{4}{9} = (5-3) + \left(\frac{}{9} - \frac{}{9}\right) = 2\frac{}{9} = 2\frac{}{3}$$

B. When subtracting mixed numbers, you may have to regroup or borrow from the whole number in order to subtract the fraction parts. Study the example carefully!

1/4 is smaller than 3/4, so borrow 1 from 6 to make 1/4 bigger.

$$6\frac{1}{4} - 3\frac{3}{4} = \left(5 + 1\frac{1}{4}\right) - 3\frac{3}{4} = 5\frac{5}{4} - 3\frac{3}{4} = 2\frac{2}{4} = 2\frac{1}{2}$$

C. You subtract mixed numbers with unlike denominators in the same way: rewrite them using a common denominator, borrow if necessary, then subtract.

8/12 is smaller than 9/12, so borrow 1 from 5.

$$5\frac{2}{3} - 1\frac{3}{4} = 5\frac{8}{12} - 1\frac{9}{12} = 4\frac{20}{12} - 1\frac{9}{12} = 3\frac{11}{12}$$

D. Subtract the mixed numbers. Simplify and convert your answers to mixed numbers.

$$3\frac{5}{6} - 2\frac{1}{6} = \qquad\qquad 5\frac{2}{9} - 1\frac{7}{9} =$$

$$8\frac{7}{9} - 5\frac{2}{3} = \qquad\qquad 4\frac{1}{3} - 2\frac{5}{6} =$$

$$3\frac{3}{4} - 2\frac{7}{8} = \qquad\qquad 5\frac{1}{6} - 3\frac{4}{5} =$$

Easy Peasy All-in-One Homeschool · EP Math 5/6 Workbook

LESSON 79 Practice

Subtract the mixed numbers. Simplify and convert your answers to mixed numbers.

$4\dfrac{7}{8} - 2\dfrac{3}{8} =$ $\qquad\qquad$ $5\dfrac{4}{9} - 4\dfrac{1}{9} =$

$4\dfrac{3}{7} - 3\dfrac{6}{7} =$ $\qquad\qquad$ $6\dfrac{5}{8} - 3\dfrac{7}{8} =$

$3\dfrac{7}{8} - 2\dfrac{1}{2} =$ $\qquad\qquad$ $4\dfrac{5}{6} - 1\dfrac{1}{3} =$

$7\dfrac{1}{3} - 4\dfrac{3}{4} =$ $\qquad\qquad$ $5\dfrac{2}{3} - 2\dfrac{8}{9} =$

$8\dfrac{1}{6} - 5\dfrac{3}{4} =$ $\qquad\qquad$ $7\dfrac{1}{2} - 4\dfrac{3}{5} =$

$9\dfrac{3}{8} - 3\dfrac{5}{6} =$ $\qquad\qquad$ $5\dfrac{1}{10} - 3\dfrac{2}{5} =$

Easy Peasy All-in-One Homeschool $\qquad\qquad$ EP Math 5/6 Workbook

LESSON 80 Word Problems: Adding/Subtracting Mixed Numbers

Solve the word problems.

Kate ran $2\frac{1}{2}$ miles yesterday. She ran $3\frac{3}{4}$ miles today. How many miles did she run in total? _____

Olivia had a ribbon $5\frac{1}{4}$ feet long. She used $1\frac{1}{2}$ feet of the ribbon to wrap a gift. How many feet of the ribbon are left? _____

David used $2\frac{5}{6}$ gallons of red paint and $3\frac{2}{3}$ gallons of white paint to paint his room. How many gallons of paint did he use in all? _____

Sarah weighs $110\frac{3}{5}$ pounds. When she holds her dog Max, she weighs $125\frac{9}{10}$ pounds. How many pounds does Max weigh? _____

Leah spent $1\frac{1}{4}$ hours studying math and 2 hours studying science. What is the total time she spent studying? _____

Andrew is $12\frac{1}{2}$ years old. His younger brother Joey is $4\frac{5}{6}$ years younger than Andrew. How old is Joey? _____

This morning $2\frac{11}{16}$ inches of snow fell. By afternoon a quarter of an inch melted away. How many inches of snow are left? _____

Mia planted $3\frac{5}{8}$ rows of roses, $4\frac{1}{2}$ rows of violets, and $5\frac{3}{4}$ rows of daisies. How many rows of flowers did Mia plant in total? _____

Kyle bought $2\frac{1}{2}$ pounds of beef and used $1\frac{7}{10}$ pounds to cook dinner. How many pounds of beef did Kyle have left after that? _____

Ana lives $2\frac{5}{8}$ miles from a park. This afternoon she walked to the park and back home. How many miles did she walk in total? _____

A restaurant had $10\frac{1}{5}$ pounds of potatoes. They used $8\frac{1}{3}$ pounds to make fries and brought in another 5 pounds. How many pounds of potatoes did the restaurant have then? _____

Easy Peasy All-in-One Homeschool

EP Math 5/6 Workbook

LESSON 80 Practice

Solve the word problems.

Matt ran $1\frac{1}{4}$ miles yesterday. He ran $3\frac{1}{2}$ miles today. How many more miles did he run today than yesterday? _____

A recipe calls for $2\frac{1}{6}$ cups of sugar for batter and $1\frac{3}{4}$ cups for sprinkling. How many cups of sugar are needed in total? _____

A tank had $9\frac{1}{4}$ gallons of water. Kyle used $4\frac{3}{5}$ gallons of water from the tank. How many gallons of water was left in the tank? _____

Brandon is $11\frac{2}{3}$ years old. His sister Sarah is $4\frac{5}{6}$ years older than Brandon. How old is Sarah? _____

Last night $3\frac{1}{4}$ inches of snow fell. This morning $2\frac{5}{8}$ inches of more snow fell. How many inches of snow are on the ground? _____

Olivia had a wire $8\frac{3}{8}$ feet long. She used $3\frac{1}{2}$ feet of the wire for her craft project. How many feet of the wire are left? _____

Jennifer used $1\frac{2}{3}$ cups of carrots and $2\frac{1}{2}$ cups of onions to make soup. How many cups of vegetables did Jennifer use in total? _____

Josh picked $4\frac{5}{6}$ baskets of apples. Tom picked $5\frac{1}{3}$ baskets of apples. How many baskets of apples did they pick all together? _____

A restaurant had $12\frac{1}{2}$ loaves of bread. They used $5\frac{7}{9}$ loaves to make sandwiches. How many loaves of bread were left? _____

A farmer decided to plant 25 rows of corns. He planted $8\frac{7}{10}$ rows yesterday and $7\frac{3}{5}$ rows today. How many rows are left? _____

Sam's cat weighs $6\frac{1}{3}$ pounds. Emily's cat weighs $1\frac{3}{4}$ pounds less than Sam's cat. Laura's cat weighs $2\frac{5}{6}$ pounds more than Emily's cat. How many pounds does Laura's cat weigh? _____

LESSON 81 Multiplying Fractions

A. To multiply fractions, multiply the numerators to find the new numerator and multiply the denominators to find the new denominator. Then simplify the result, if possible. Here is an example of using this method. Complete the example.

$$\frac{5}{8} \times \frac{4}{9} = \frac{5 \times 4}{8 \times 9} = \frac{}{72} = \frac{}{18}$$

B. Another way to multiply fractions is to simplify them using cross reducing before multiplying. Cross reducing means dividing the numerator of one fraction and the denominator of the other fraction by a common factor. This method is easier and faster because it helps you avoid dealing with large numbers. See the examples below.

$$\frac{5 \times 4}{8 \times 9} = \frac{\cancel{4} \times 5}{\cancel{8} \times 9} = \frac{1 \times 5}{2 \times 9} = \frac{5}{18}$$ Cross reducing is the same as performing these steps, but it is faster!

$$\frac{5}{8} \times \frac{4}{9} \Rightarrow$$ 4 and 8 have a common factor 4, so cross reduce. $$\Rightarrow \frac{5}{\cancel{8}2} \times \frac{\cancel{4}1}{9} = \frac{5}{18}$$

$$\frac{5}{6} \times \frac{4}{15} \Rightarrow$$ 5 and 15 have a common factor. So do 4 and 6. $$\Rightarrow \frac{\cancel{5}1}{\cancel{6}3} \times \frac{\cancel{4}2}{\cancel{15}3} = \frac{2}{9}$$

C. Multiply the fractions. Simplify and convert your answers to mixed numbers.

$$\frac{1}{2} \times \frac{4}{5} = \qquad\qquad \frac{8}{12} \times \frac{9}{10} =$$

$$\frac{3}{8} \times \frac{2}{9} = \qquad\qquad \frac{9}{24} \times \frac{6}{18} =$$

$$\frac{4}{9} \times \frac{6}{8} = \qquad\qquad \frac{10}{21} \times \frac{14}{15} =$$

Easy Peasy All-in-One Homeschool — EP Math 5/6 Workbook

LESSON 81 Practice

Multiply the fractions. Simplify and convert your answers to mixed numbers.

$\dfrac{2}{3} \times \dfrac{1}{4} =$ $\qquad\qquad$ $\dfrac{8}{22} \times \dfrac{2}{16} =$

$\dfrac{3}{5} \times \dfrac{5}{7} =$ $\qquad\qquad$ $\dfrac{7}{12} \times \dfrac{6}{14} =$

$\dfrac{2}{6} \times \dfrac{3}{4} =$ $\qquad\qquad$ $\dfrac{6}{13} \times \dfrac{13}{15} =$

$\dfrac{2}{9} \times \dfrac{5}{6} =$ $\qquad\qquad$ $\dfrac{5}{54} \times \dfrac{18}{40} =$

$\dfrac{3}{8} \times \dfrac{2}{5} =$ $\qquad\qquad$ $\dfrac{15}{16} \times \dfrac{12}{18} =$

$\dfrac{6}{9} \times \dfrac{3}{4} =$ $\qquad\qquad$ $\dfrac{27}{50} \times \dfrac{25}{60} =$

Easy Peasy All-in-One Homeschool $\qquad\qquad$ EP Math 5/6 Workbook

Lesson 82 Multiplying Mixed Numbers

A. To multiply mixed numbers, convert them to improper fractions and multiply the improper fractions. Then simplify the result and convert it back to a mixed number, if possible. Here is an example of using this method. Complete the example.

$$5\frac{1}{4} \times 3\frac{5}{9} = \frac{21}{4} \times \frac{32}{9} = \frac{\cancel{21}\,7}{\cancel{4}\,1} \times \frac{\cancel{32}\,8}{\cancel{9}\,3} = \frac{}{3} = 18\frac{}{3}$$

B. Multiply the fractions. Simplify and convert your answers to mixed numbers.

$1\dfrac{1}{2} \times 1\dfrac{1}{9} =$ $\qquad\qquad$ $2\dfrac{2}{5} \times 2\dfrac{1}{12} =$

$2\dfrac{1}{4} \times 1\dfrac{5}{6} =$ $\qquad\qquad$ $2\dfrac{1}{7} \times 1\dfrac{2}{15} =$

$2\dfrac{1}{3} \times 3\dfrac{3}{7} =$ $\qquad\qquad$ $2\dfrac{5}{8} \times 1\dfrac{1}{15} =$

$5\dfrac{5}{9} \times 4\dfrac{4}{5} =$ $\qquad\qquad$ $2\dfrac{2}{3} \times 2\dfrac{7}{10} =$

$2\dfrac{5}{8} \times 5\dfrac{1}{3} =$ $\qquad\qquad$ $3\dfrac{3}{7} \times 1\dfrac{1}{8} =$

Easy Peasy All-in-One Homeschool $\qquad\qquad$ EP Math 5/6 Workbook

LESSON 82 Practice

Multiply the fractions. Simplify and convert your answers to mixed numbers.

$3\dfrac{3}{8} \times 1\dfrac{1}{3} =$ $\qquad$ $3\dfrac{1}{3} \times 1\dfrac{7}{10} =$

$1\dfrac{3}{5} \times 2\dfrac{1}{8} =$ $\qquad$ $4\dfrac{3}{8} \times 1\dfrac{5}{21} =$

$5\dfrac{2}{5} \times 1\dfrac{1}{3} =$ $\qquad$ $4\dfrac{1}{6} \times 1\dfrac{3}{15} =$

$1\dfrac{3}{4} \times 2\dfrac{2}{7} =$ $\qquad$ $3\dfrac{1}{7} \times 3\dfrac{6}{11} =$

$2\dfrac{5}{8} \times 1\dfrac{3}{9} =$ $\qquad$ $1\dfrac{2}{3} \times 1\dfrac{7}{20} =$

$3\dfrac{1}{7} \times 1\dfrac{3}{4} =$ $\qquad$ $6\dfrac{1}{8} \times 1\dfrac{4}{14} =$

Easy Peasy All-in-One Homeschool EP Math 5/6 Workbook

LESSON 83 Multiplying Fractions of All Types

A. To multiply fractions with whole numbers, rewrite the whole numbers as fractions and multiply the fractions as usual. To rewrite a whole number as a fraction, simply put the whole number over 1. Complete the example.

$$\frac{7}{12} \times 15 = \frac{7}{\cancel{12}\,4} \times \frac{\cancel{15}\,5}{1} = \frac{}{4} = 8\frac{}{4}$$

B. When multiplying fractions, including mixed numbers and whole numbers, remember to rewrite any mixed number as an improper fraction and any whole number as a fraction before multiplying. Complete the examples.

$$\frac{4}{9} \times 3\frac{3}{8} = \frac{\cancel{4}\,1}{\cancel{9}\,1} \times \frac{\cancel{27}\,3}{\cancel{8}\,2} = \frac{}{2} = 1\frac{}{2}$$

$$2\frac{7}{9} \times 12 = \frac{25}{9} \times 12 = \frac{25}{\cancel{9}\,3} \times \frac{\cancel{12}\,4}{1} = \frac{}{3} = 33\frac{}{3}$$

C. Multiply the fractions. Don't forget to reduce first before multiplying. Simplify and convert your answers to mixed numbers.

$\frac{3}{8} \times 4 =$ $\qquad$ $\frac{2}{5} \times 1\frac{8}{7} =$

$7 \times \frac{2}{5} =$ $\qquad$ $3\frac{1}{2} \times \frac{4}{21} =$

$\frac{5}{9} \times 6 =$ $\qquad$ $16 \times 2\frac{5}{6} =$

$8 \times \frac{2}{3} =$ $\qquad$ $20 \times 1\frac{2}{8} =$

Easy Peasy All-in-One Homeschool $\qquad$ EP Math 5/6 Workbook

LESSON 83 Practice

Multiply the fractions. Simplify and convert your answers to mixed numbers.

$\dfrac{1}{2} \times 4 =$

$\dfrac{4}{7} \times 1\dfrac{5}{8} =$

$9 \times \dfrac{1}{6} =$

$3\dfrac{1}{9} \times \dfrac{3}{4} =$

$\dfrac{3}{4} \times 6 =$

$\dfrac{5}{18} \times 4\dfrac{1}{5} =$

$\dfrac{5}{6} \times 8 =$

$1\dfrac{1}{9} \times 15 =$

$8 \times \dfrac{5}{8} =$

$8 \times 1\dfrac{5}{12} =$

$7 \times \dfrac{7}{9} =$

$2\dfrac{3}{20} \times 6 =$

Easy Peasy All-in-One Homeschool

EP Math 5/6 Workbook

LESSON 84 Word Problems: Multiplying Fractions

A. Multiplying a fraction means finding a part of something. Multiplying a fraction by a fraction means finding a part of a part of something. The following examples show this visually. Complete the examples. The word "of" is your clue to multiply.

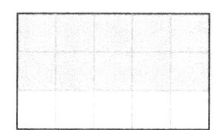

How many are 3/4 of 12 dots? $\Rightarrow$ $\dfrac{3}{4} \times 12 = \dfrac{3}{\cancel{4}\,1} \times \dfrac{\cancel{12}\,3}{1} = \underline{}$ dots

2/5 of 2/3 of the rectangle is colored. What fraction of the rectangle is colored? $\Rightarrow$ $\dfrac{2}{5} \times \dfrac{2}{3} = \dfrac{}{15}$

B. Solve. Make sure to not just multiply but check to see what question is being asked.

Tylor had 9 candies and ate $\dfrac{1}{3}$ of them. Lucy had 8 candies and ate $\dfrac{1}{4}$ of them. Who ate more candies? _____

Kelly spent $2\dfrac{5}{6}$ hours studying math. She spent $\dfrac{3}{5}$ of that time on fractions. How many hours did Kelly spend on fractions? _____

In a class, $\dfrac{3}{5}$ of the students have pets. Among those students, $\dfrac{2}{3}$ have dogs. What fraction of the students in the class have dogs? _____

Two and three-quarters pizzas were in the fridge. Emily and her brother ate $\dfrac{2}{3}$ of them. How much pizza was left in the fridge? _____

Justin had $18 and spent $\dfrac{5}{9}$ of his money. Sandy had $25 and spent $\dfrac{3}{5}$ of her money. Who has more money left? _____

The bucket had $5\dfrac{1}{4}$ gallons of water. Matt used $\dfrac{1}{3}$ of that water to water his plants and added $2\dfrac{1}{2}$ gallons of water to the bucket. How many gallons of water were in the bucket after that? _____

A restaurant had 25 pounds of potatoes. They used $\dfrac{3}{5}$ of the potatoes to make fries and the rest to make potato salad. How many pounds of potatoes were used to make potato salad? _____

Easy Peasy All-in-One Homeschool   EP Math 5/6 Workbook

LESSON 84 Practice

Solve. Make sure to not just multiply but check to see what question is being asked.

Walter had $16 and spent $\frac{7}{8}$ of his money. Lisa had $36 and spent $\frac{3}{4}$ of her money. Who spent more money? _____

At a pet store, $\frac{1}{5}$ of the pets are birds. Of those birds, $\frac{2}{3}$ are male. What fraction of the pets in the store are male birds? _____

Clay spent $\frac{4}{7}$ of his money on books. Of that money, $\frac{5}{8}$ was spent on novels. What fraction of Clay's money was spent on novels? _____

Carlos solved 20 fraction problems. He got $\frac{3}{4}$ of the problems correct. How many problems did he get incorrect? _____

Steve ran $2\frac{1}{2}$ miles every day for the past $3\frac{1}{7}$ weeks. How many miles did Steve run in total? _____

Jim saved $\frac{5}{8}$ of his allowance. One fifth of that money was for a bike. What fraction of Jim's allowance was for the bike? _____

Ellen solved 25 problems and got $\frac{4}{5}$ of the problems correct. She solved the incorrect ones again and got three fifths of them correct. How many problems did she get incorrect after all? _____

A recipe calls for $1\frac{3}{4}$ cups of sugar for one pie, and $\frac{1}{2}$ of the sugar is for sprinkling after baking. Jessica wants to bake five pies. How many cups of sugar will she need for sprinkling? _____

A pet store had 125 goldfish. They sold $\frac{4}{5}$ of the goldfish and brought in another 150 goldfish. How many goldfish did the store have then? _____

Rodney used $4\frac{2}{5}$ gallons of paint to paint the fence. Five elevenths of the paint was yellow and the rest was green. How many gallons of green paint did Rodney use? _____

Easy Peasy All-in-One Homeschool   EP Math 5/6 Workbook

LESSON 85 Dividing Fractions

A. To divide fractions, rewrite division as multiplication by multiplying by the reciprocal of the divisor. Reduce and then multiply the fractions as usual. Simplify the result and convert it to a mixed number, if possible. The reciprocal of a fraction is the fraction turned upside down. Here is an example of using this method. Complete the example.

$$\frac{8}{15} \div \frac{4}{25} = \frac{\cancel{8}2}{\cancel{15}3} \times \frac{\cancel{25}5}{\cancel{4}1} = \frac{}{3} = 3\frac{}{3}$$

B. A whole number can be expressed as a fraction by placing the whole number over 1. The reciprocal of a whole number (except 0) is 1 over the number. Here are some examples of rewriting whole numbers as fractions before dividing. Complete the examples.

$$\frac{7}{12} \div 21 = \frac{7}{12} \div \frac{21}{1} = \frac{\cancel{7}1}{12} \times \frac{1}{\cancel{21}3} = \frac{}{36}$$

$$27 \div \frac{6}{11} = \frac{27}{1} \div \frac{6}{11} = \frac{\cancel{27}9}{1} \times \frac{11}{\cancel{6}2} = \frac{}{2} = 49\frac{}{2}$$

C. Divide the fractions. Simplify and convert your answers to mixed numbers.

$$\frac{5}{6} \div \frac{3}{4} = \qquad\qquad \frac{11}{12} \div \frac{33}{45} =$$

$$8 \div \frac{6}{7} = \qquad\qquad 16 \div \frac{48}{49} =$$

$$\frac{3}{5} \div 9 = \qquad\qquad \frac{10}{14} \div 35 =$$

Easy Peasy All-in-One Homeschool   EP Math 5/6 Workbook

LESSON 85 Practice

Divide the fractions. Simplify and convert your answers to mixed numbers.

$\dfrac{2}{3} \div \dfrac{1}{4} =$

$\dfrac{9}{14} \div \dfrac{18}{32} =$

$\dfrac{1}{2} \div \dfrac{5}{6} =$

$\dfrac{12}{13} \div \dfrac{20}{39} =$

$4 \div \dfrac{8}{9} =$

$10 \div \dfrac{15}{18} =$

$6 \div \dfrac{6}{7} =$

$25 \div \dfrac{12}{15} =$

$\dfrac{3}{5} \div 9 =$

$\dfrac{16}{40} \div 24 =$

$\dfrac{4}{7} \div 2 =$

$\dfrac{12}{27} \div 36 =$

Easy Peasy All-in-One Homeschool   EP Math 5/6 Workbook

LESSON 86 Word Problems: Dividing Fractions

A. Division is the inverse of multiplication. Dividing by a number is the same as multiplying by its reciprocal or multiplicative inverse. Consider the example below.

Divide 12 candies equally into 4 boxes. $= 12 \div 4 = 12 \times \dfrac{1}{4} =$ Put 1/4 of 12 candies into each box.

B. Dividing by a fraction means finding how many of the fractional parts are in a whole. Follow the examples below to understand what it means in real life.

How many half hours are in 4 hours? $= 4 \div \dfrac{1}{2} = \dfrac{4}{1} \times \dfrac{2}{1} = 8 =$ There are 8 half hours in 4 hours.

How many quarter inches are in 1/2 inches? $= \dfrac{1}{2} \div \dfrac{1}{4} = \dfrac{1}{2} \times \dfrac{4}{1} = 2 =$ There are 2 quarter inches in 1/2 inches.

C. Solve the word problems.

Lucas did a total of 4 hours of fitness training, which consisted of half hour sessions. How many sessions did he do? _____

Mia had a $\dfrac{7}{8}$-inch ribbon. For her craft project, she cut it into quarter-inch pieces. How many pieces did she have in total? _____

Mark had $\dfrac{9}{10}$ gallon of milk. He divided the milk evenly into 3 bottles. How much milk did he put in each bottle? _____

Elijah took a walk for $\dfrac{3}{4}$ hour. It usually takes him $\dfrac{1}{3}$ hour to walk a mile. How many miles did he walk? _____

Charlie takes $\dfrac{5}{6}$ minute to solve one fraction problem. How many problems can he solve in 30 minutes? _____

Brian used $\dfrac{7}{16}$ gallon of red paint and $\dfrac{7}{8}$ gallon of blue paint to paint 3 bookshelves. How many gallons of paint did he use for each bookshelf? _____

A math riddle for you! Find the smallest number that is divisible by all the numbers 1 to 10.

Easy Peasy All-in-One Homeschool · EP Math 5/6 Workbook

LESSON 86 Practice

Solve the word problems.

Paul takes $\frac{1}{6}$ hour to read a page of his reading assignment. How many pages can he read in $\frac{2}{3}$ hour? _____

A cook used $\frac{5}{6}$ cup of flour to bake 5 cookies. How many cups of flour did the cook use to make one cookie? _____

Sara has $\frac{3}{4}$ of a cake. She shares it equally among herself and 2 friends. What fraction of the cake will each get? _____

Amy has 7 gallons of gas in her car. Her car uses $\frac{1}{25}$ of a gallon per mile. How far can Amy drive before the tank is empty? _____

A recipe calls for $\frac{5}{8}$ cup of flour to make a pie. Noah has 10 cups of flour. How many pies will he be able to make? _____

Debora made 6 gallons of lemonade. Each cup holds $\frac{1}{12}$ of a gallon. How many cups will Debora be able to fill? _____

A farmer plans to plant four different vegetables in $\frac{6}{7}$ of his field. What fraction of the field will be occupied by each vegetable? _____

Aurora had a $\frac{15}{16}$-inch wire. For her science project, she cut it into $\frac{1}{8}$-inch pieces. How many pieces did she have? _____

Mateo can run 5 laps in $\frac{1}{4}$ hour. How many <u>minutes</u> does it take him to run a lap? _____

Mike wants to put $\frac{3}{4}$-inch square tiles on a 12-inch square board. How many rows and columns of tiles will he need? How many tiles will he use in all? _____

A recipe calls for $\frac{7}{8}$ cup of diced vegetables to make 4 servings of soup. Sally wants to make 6 servings. How many cups of vegetables does she need? _____

Easy Peasy All-in-One Homeschool — EP Math 5/6 Workbook

LESSON 87 Reciprocal of a Mixed Number

A. To find the reciprocal of a fraction, just turn the fraction upside down. The reciprocal of a whole number (except 0) is 1 over the number. The reciprocal of a unit fraction, a fraction whose numerator is 1, is a whole number. Write the reciprocal of each fraction.

$\dfrac{7}{12}$ _____   18 _____ $\dfrac{20}{9}$ _____ $\dfrac{1}{45}$ _____

B. To find the reciprocal of a mixed number, convert it to an improper fraction and turn the improper fraction upside down. Write the reciprocal of each mixed number.

$3\dfrac{1}{2}$ _____ $4\dfrac{1}{5}$ _____ $3\dfrac{3}{5}$ _____ $1\dfrac{7}{10}$ _____

$6\dfrac{2}{7}$ _____ $1\dfrac{4}{9}$ _____ $2\dfrac{5}{6}$ _____ $3\dfrac{4}{15}$ _____

C. Find the reciprocal of each fraction. Write your answers as mixed numbers, if possible.

$\dfrac{17}{6}$ _____ $\dfrac{1}{18}$ _____ $\dfrac{13}{35}$ _____   16 _____

20 _____ $\dfrac{21}{4}$ _____ $\dfrac{1}{30}$ _____ $\dfrac{11}{25}$ _____

D. What fraction am I? Use the clues to find the correct fractions.

My numerator is a factor of all numbers. The sum of myself and my reciprocal is five and one fifth. What fraction am I?   _____

My denominator is 6 more than my numerator. The simplest form of my reciprocal is two and one half. What fraction am I?   _____

The denominator of my reciprocal is the smallest even prime number. My denominator is 5 more than my numerator. What fraction am I?   _____

Did you know? 2 and 5 are the only prime numbers that end in 2 or 5.

Easy Peasy All-in-One Homeschool   EP Math 5/6 Workbook

LESSON 87 Practice

A. Find the reciprocal of each fraction. Write your answers as mixed numbers, if possible.

$\dfrac{1}{25}$ _____ $2\dfrac{7}{10}$ _____ 40 _____ $\dfrac{12}{25}$ _____

26 _____ $3\dfrac{6}{13}$ _____ $\dfrac{1}{23}$ _____ $5\dfrac{2}{7}$ _____

$3\dfrac{2}{3}$ _____ $\dfrac{1}{48}$ _____ $\dfrac{21}{32}$ _____ 52 _____

$\dfrac{15}{49}$ _____ 35 _____ $3\dfrac{7}{8}$ _____ $\dfrac{1}{19}$ _____

B. What fraction am I? Use the clues to find the correct fractions.

I am the sum of $1\dfrac{1}{2}$ and its reciprocal. What fraction am I? _____

I am the difference of $1\dfrac{1}{5}$ and its reciprocal. What fraction am I? _____

I am equivalent to $2\dfrac{1}{2}$ and my numerator is 5^2. What fraction am I? _____

My denominator is 2 more than my numerator. The simplest form of my reciprocal is one and one third. What fraction am I? _____

My denominator is 3 less than my numerator. My reciprocal is equivalent to three fourths. What fraction am I? _____

My numerator is 3 less than my denominator. If both my numerator and my denominators are increased by 2, I become equivalent to four fifths. What fraction am I? _____

I am a proper fraction. My numerator and my denominator are prime numbers. The least common denominator of myself and my reciprocal is 6. What fraction am I? _____

LESSON 88 Dividing Fractions of All Types

A. To divide mixed numbers, convert them to improper fractions and divide the improper fractions. Reduce and then simplify the result and convert it back to a mixed number, if possible. Here is an example of using this method. Complete the example.

$$5\frac{1}{4} \div 3\frac{1}{9} = \frac{21}{4} \div \frac{28}{9} = \frac{\cancel{21}\,3}{4} \times \frac{9}{\cancel{28}\,4} = \frac{}{16} = 1\frac{}{16}$$

B. Remember to rewrite any mixed number as an improper fraction and any whole number as a fraction before dividing. Complete the examples.

$$\frac{7}{12} \div 21 = \frac{7}{12} \div \frac{21}{1} = \frac{\cancel{7}\,1}{12} \times \frac{1}{\cancel{21}\,3} = \frac{}{36}$$

$$12 \div 3\frac{1}{5} = \frac{12}{1} \div \frac{16}{5} = \frac{\cancel{12}\,3}{1} \times \frac{5}{\cancel{16}\,4} = \frac{}{4} = 3\frac{}{4}$$

C. Divide the fractions. Simplify and convert your answers to mixed numbers. Before you multiply, check to see if you can reduce.

$$\frac{4}{9} \div 16 = \qquad\qquad 1\frac{2}{3} \div 30 =$$

$$20 \div \frac{15}{21} = \qquad\qquad 12 \div 3\frac{3}{4} =$$

$$\frac{5}{7} \div 4\frac{1}{6} = \qquad\qquad 4\frac{2}{7} \div 3\frac{3}{4} =$$

$$2\frac{4}{5} \div \frac{7}{10} = \qquad\qquad 4\frac{1}{6} \div 3\frac{4}{7} =$$

Easy Peasy All-in-One Homeschool EP Math 5/6 Workbook

LESSON 88 Practice

Divide the fractions. Simplify and convert your answers to mixed numbers.

$9 \div \dfrac{3}{4} =$

$14 \div 1\dfrac{1}{6} =$

$\dfrac{3}{8} \div 12 =$

$5\dfrac{1}{4} \div 18 =$

$\dfrac{2}{9} \div 1\dfrac{2}{7} =$

$8\dfrac{3}{4} \div 2\dfrac{1}{7} =$

$2\dfrac{4}{7} \div \dfrac{3}{7} =$

$9\dfrac{1}{3} \div 2\dfrac{4}{5} =$

$\dfrac{3}{5} \div 1\dfrac{4}{5} =$

$3\dfrac{1}{5} \div 6\dfrac{2}{3} =$

$3\dfrac{3}{7} \div \dfrac{6}{11} =$

$4\dfrac{3}{8} \div 1\dfrac{2}{5} =$

Easy Peasy All-in-One Homeschool

EP Math 5/6 Workbook

LESSON 89 Decimal Place Value

A. Decimals are numbers that contain a decimal point. Just like whole numbers, each digit in a decimal number has a place value 10 times the place value of the digit to its right. The decimal point is placed after the ones digit to separate the whole number part from the decimal part. The chart below shows place values for decimals. Complete the chart.

	Thousands	Hundreds	Tens	Ones	Decimal Point	Tenths	Hundredths	Thousandths
	1000	100	10	1		$\frac{1}{10}$	$\frac{1}{100}$	$\frac{1}{1000}$
241.82 =	-	2	4	1	.	8	2	-
1,032.497 =								

B. To read and write decimals, read the digits before the decimal point as a whole number, say "and" for the decimal point, read the digits after the decimal point as a whole number, and then say the place value of the last digit. Here are some examples. Note that there's no need to say "zero and." Say each number out loud.

12.3 twelve and three tenths

438.261 four hundred thirty-eight and two hundred sixty-one thousandths

0.06 six hundredths

18,000.5 eighteen thousand and five tenths

C. Write the place value of the underlined digit in each number.

32.<u>4</u> **tenths**   <u>6</u>2.35   17.3<u>5</u>

<u>2</u>.47   2.<u>9</u>13   0.23<u>8</u>

D. Write each number in word form.

70.35 =

0.217 =

Easy Peasy All-in-One Homeschool   EP Math 5/6 Workbook

LESSON 89 Practice

A. Make a number using each set of place values.

7 tens	5 hundreds	6 thousands	3 tenths
1 one	2 tens	3 hundreds	4 tens
4 tenths	0 ones	4 tens	8 thousandths
3 hundredths	8 tenths	9 ones	2 hundredths
5 thousandths	4 hundredths	2 tenths	6 ones

_____   _____   _____   _____

B. Write the place value of the underlined digit in each number.

1̲7.3   23.1̲5   0.527̲

2.6̲1   5̲.434   12.३5̲

C. Write each number in word form.

215.4 =

0.639 =

9,015.02 =

D. What number am I? Use the clues to find the decimal (with a decimal point).

I have 2 digits in total. If you multiply any number by my tenths digit, you get the number itself. My ones digit is the square of my tenths digit. What number am I? _____

I have 3 digits in total. If you multiply any number by my ones digit, you get zero. My tens digit is the largest single digit. The sum of my digits is 18. What number am I? _____

I have 3 digits in total. All of my digits are even. My ones digit is twice my tenths digit. My tenths digit is twice my hundredths digit. What number am I? _____

Easy Peasy All-in-One Homeschool   EP Math 5/6 Workbook

LESSON 90 Decimals on a Number Line

A. To place decimals with tenths on a number line, divide intervals between two successive integers into 10 equal parts. To place decimals with hundredths, divide intervals of one-tenth into 10 equal parts. To place decimals with thousandths, divide intervals of one-hundredth into 10 equal parts. Here are some examples of decimals placed on number lines. Fill in the missing decimals.

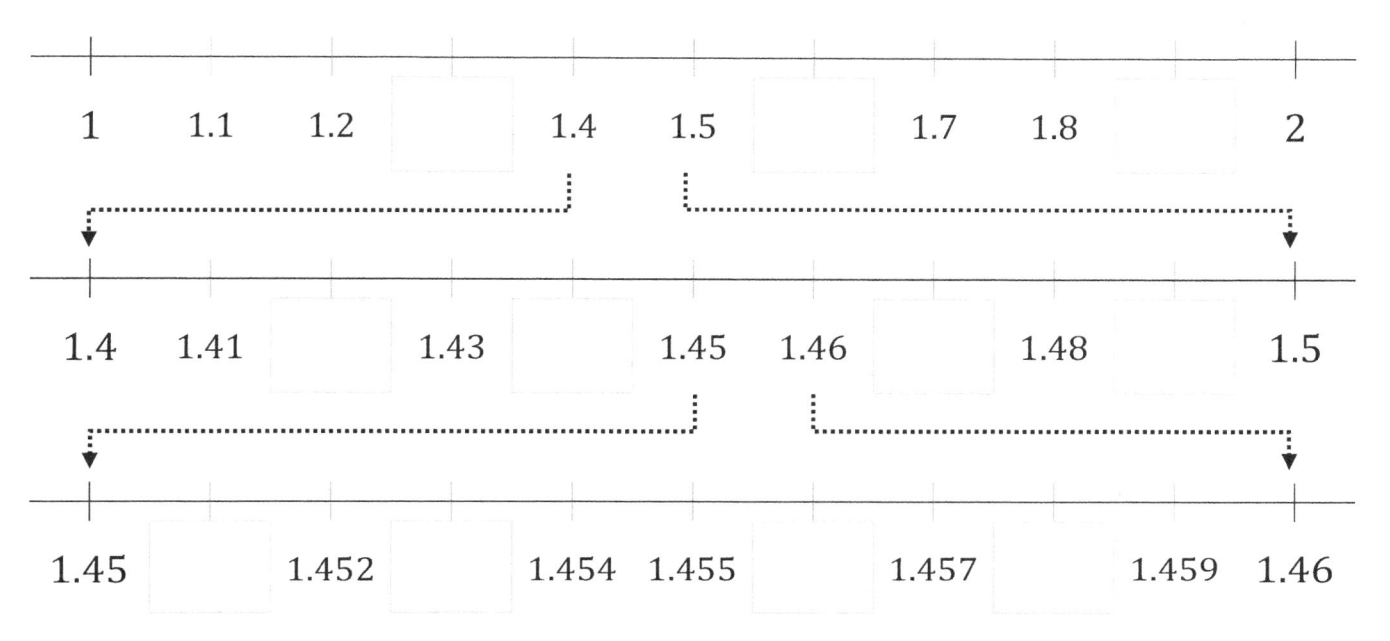

1   1.1   1.2   ☐   1.4   1.5   ☐   1.7   1.8   2

1.4   1.41   ☐   1.43   ☐   1.45   1.46   ☐   1.48   1.5

1.45   ☐   1.452   ☐   1.454   1.455   ☐   1.457   ☐   1.459   1.46

B. Write the whole number and decimal value of each letter marked on the number lines.

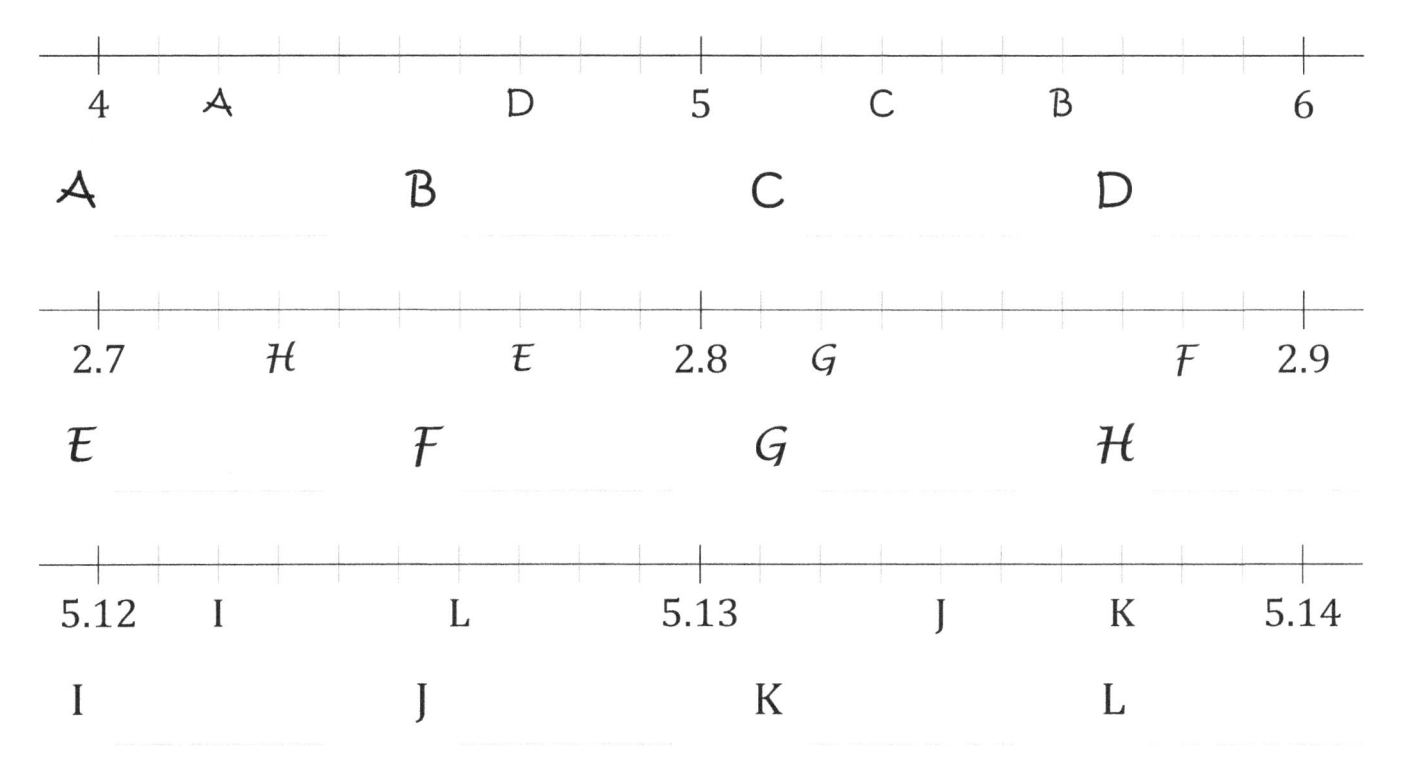

4   A   D   5   C   B   6
A   B   C   D

2.7   H   E   2.8   G   F   2.9
E   F   G   H

5.12   I   L   5.13   J   K   5.14
I   J   K   L

Easy Peasy All-in-One Homeschool   EP Math 5/6 Workbook

LESSON 90 Practice

A. Fill in the missing decimals.

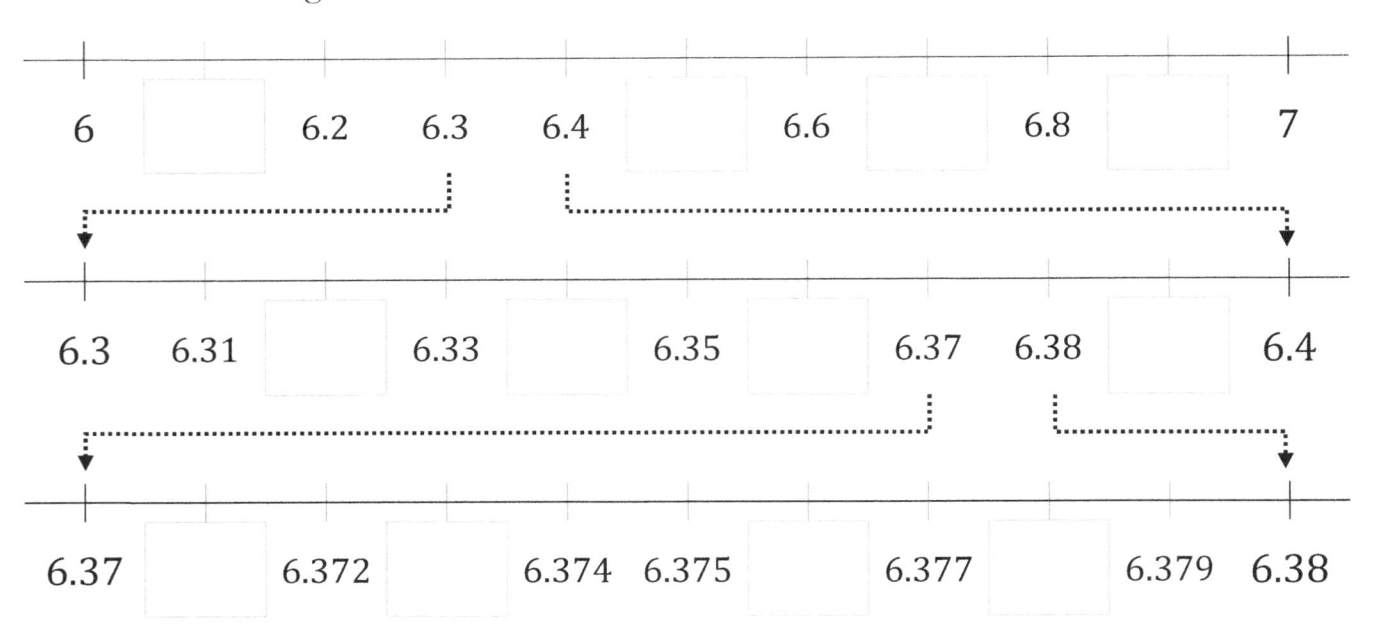

```
6    [6.1]   6.2   6.3   6.4   [6.5]   6.6   [6.7]   6.8   [6.9]   7
```

```
6.3   6.31   [6.32]   6.33   [6.34]   6.35   [6.36]   6.37   6.38   [6.39]   6.4
```

```
6.37   [6.371]   6.372   [6.373]   6.374   6.375   [6.376]   6.377   [6.378]   6.379   6.38
```

B. Write the decimal value of each letter marked on the number lines.

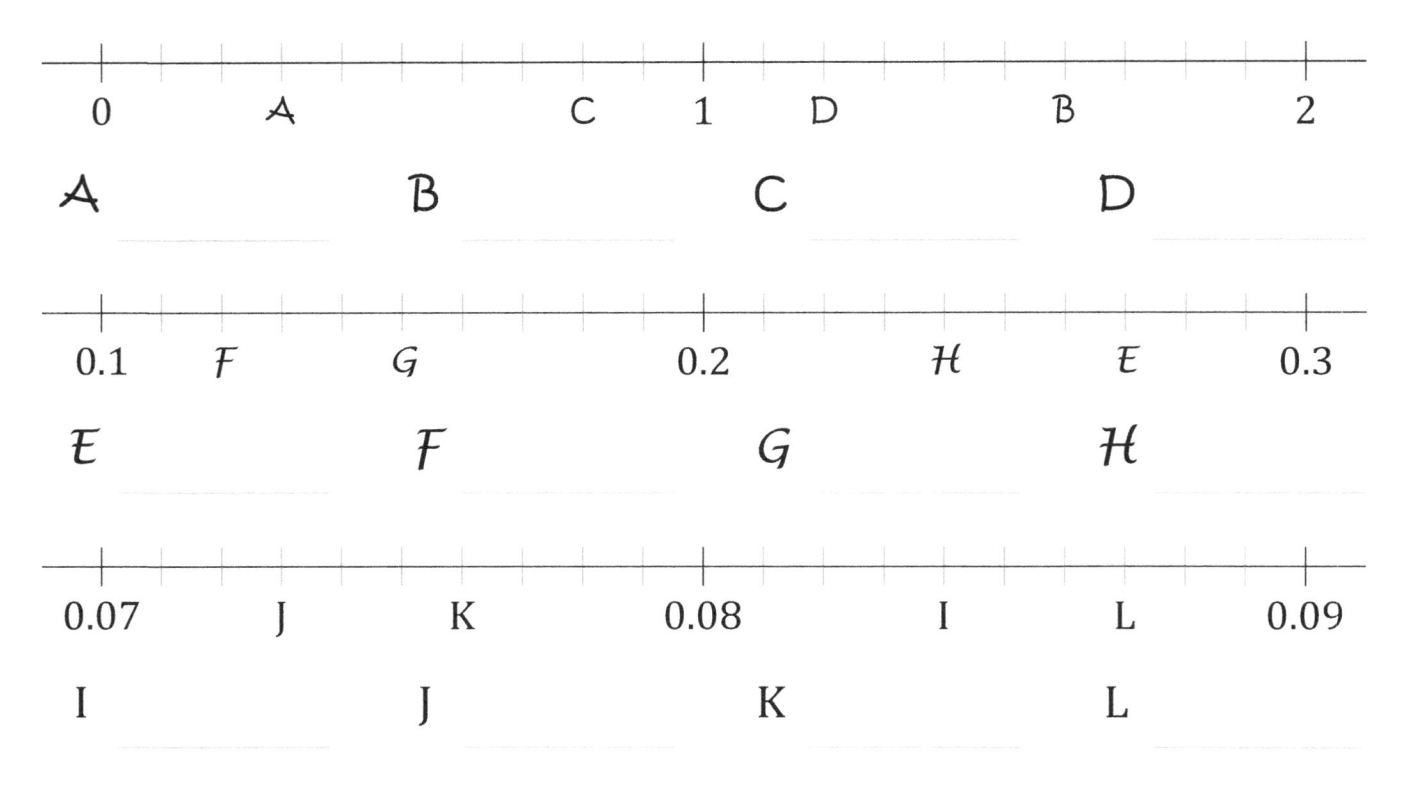

```
0    A         C    1    D         B    2
```

A _____   B _____   C _____   D _____

```
0.1   F   G         0.2         H         E   0.3
```

E _____   F _____   G _____   H _____

```
0.07   J   K         0.08         I   L   0.09
```

I _____   J _____   K _____   L _____

A math riddle for you! Jessica and her younger brother Jake are jogging. To jog alongside, Jake requires three steps to match every two steps of Jessica. If both start with their right foot, after how many steps will their left feet be together?

Easy Peasy All-in-One Homeschool   EP Math 5/6 Workbook

LESSON 91 Rounding Decimals

A. Rounding decimals is similar to rounding whole numbers except for one thing: you drop the digits to the right of the given place value instead of replacing them with zeros. Here is an example of rounding a decimal to various place values.

	Nearest whole number	Nearest tenth	Nearest hundredth
8.547	9	8.5	8.55
	The tenths place is 5. Round up and drop 547.	The hundredths place is 4. Round down and drop 47.	The thousandths place is 7. Round up and drop 7.

B. Consider the questions below and compare your answers with Part **A**.

6.547

Is 6.5 closer to 6 or 7?   Is 6.54 closer to 6.5 or 6.6?   Is 6.547 closer to 6.54 or 6.55?

C. Round each number to the nearest whole number, tenth, and hundredth.

	Nearest whole number	Nearest tenth	Nearest hundredth
0.538			
6.765			
2.203			
18.397			
64.055			
30.736			
819.542			
230.609			
1,378.003			

LESSON 91 Practice

A. Round each number to the nearest whole number.

0.2 ____	1.57 ____	1.939 ____
19.8 ____	43.26 ____	25.088 ____
353.7 ____	179.92 ____	960.507 ____

B. Round each number to the nearest tenth.

0.07 ____	5.02 ____	1.769 ____
24.39 ____	13.25 ____	80.204 ____
569.72 ____	294.99 ____	343.556 ____

C. Round each number to the nearest hundredth.

0.008 ____	2.135 ____	5.705 ____
0.051 ____	36.759 ____	23.091 ____
0.967 ____	120.022 ____	458.399 ____

D. Use the clues to find the correct decimal from the six possibilities.

7.09   4.64   5.18   6.23   4.94   5.02

To the nearest whole number, I round to 5. My tenths digit is odd. I round down to the nearest tenth. What number am I? _____

To the nearest ten, I round to 10. My tenths digit is neither positive nor negative. I round up to the nearest tenth. What number am I? _____

I round down to the nearest whole number. I round down to the nearest tenth. On a number line, I am left of 6. What number am I? _____

Easy Peasy All-in-One Homeschool   EP Math 5/6 Workbook

LESSON 92 Estimation with Decimals

A. Rounding is used when we don't need an exact value and an estimate is good enough. In real life, we often round decimals to the nearest whole number to estimate sums and differences. The following example shows one such situation. Complete the example.

Mia saved $26.75 last month and $38.16 this month. About how much money did she save in total? _____

Estimate each amount to the nearest dollar:	Find the estimated sum:
Last month: $26.75 is about $27.	$27 + $38 =
This month: $38.16 is about $38.	

B. Estimate the answers by rounding each decimal to the nearest whole number.

$8.2 + 6.7 = 8 + 7 = 15$ $9.3 - 2.7 =$

$8.54 + 9.10 =$ $6.45 - 3.52 =$

$92.48 + 63.05 =$ $84.27 - 26.61 =$

$25.12 + 39.76 =$ $43.35 - 19.88 =$

C. Solve the word problems.

A video game costs $27.99. It's on sale for $19.99. About how much less is the sale price than the original price? _____

At a book store, Graham bought three books at $9.20, $11.65, and $14.80. About how much money did Graham spend in all? _____

Last night 3.7 inches of snow fell. This morning 2.25 inches of more snow fell. About how much snow is on the ground? _____

Dallas, Texas, receives 2.1 inches of rainfall on average in January. In February, it receives 2.6 inches on average. About how much rainfall does Dallas receive during these two months? _____

A lion can run up to 80.5 kilometers per hour. A cheetah can run 120.7 kilometers per hour. About how much faster can the cheetah run than the lion? _____

LESSON 92 Practice

A. Estimate the sums by rounding each decimal to the nearest whole number.

$9.2 + 4.5 =$　　　　　　　　　　$6.9 + 3.2 =$

$18.7 + 25.4 =$　　　　　　　　　$45.4 + 20.9 =$

$56.26 + 20.13 =$　　　　　　　　$67.68 + 11.55 =$

B. Estimate the differences by rounding each decimal to the nearest whole number.

$9.1 - 4.5 =$　　　　　　　　　　$8.9 - 5.3 =$

$50.2 - 25.5 =$　　　　　　　　　$68.6 - 49.7 =$

$37.53 - 12.76 =$　　　　　　　　$79.09 - 30.82 =$

C. Solve the word problems.

Adam is buying a math puzzle book for $8.50. If he pays with a $20 bill, about how much change will he get back? _____

Lisa wants to buy a hat for $12.44 and a shirt for $23.60. About how much money will she need to buy both? _____

This morning 2.3 inches of snow fell. By afternoon 0.9 inch of snow melted away. About how much snow is left? _____

Ana used 4.8 gallons of red paint and 3.6 gallons of white paint to paint her room. About how much paint did she use in all? _____

David went for a run on a 5.2-mile trail. He ran 3.8 miles and walked the rest. About how many miles did he walk? _____

At a bookstore, Jamie bought three books at $11.99, $7.45, and $12.55. About how much money did Jamie spend in all? _____

Mercury travels around the sun at 47.87 kilometers per second. Earth travels at 29.78 kilometers per second. About how much faster does Mercury travel than Earth? _____

LESSON 93 Comparing Decimals

A. Adding zeros to the end of a decimal number does not change its value. Note that the following decimals all have the same value.

$$0.3 = 0.30 = 0.300$$ 3 tenths = 30 hundredths = 300 thousandths

B. Here are the steps for comparing decimals. Complete the example.

(1) 57.36   57.3

(2) 57.36 ☐ 57.30

To compare 57.36 to 57.3:
1. Compare the whole number parts. Both are 57.
2. Compare the decimal parts. If the decimal parts have different lengths (different numbers of digits), add zeros to the end of the shorter decimal so they have the same length. Add one zero to the end of 57.3 and compare 36 to 30.

C. Compare the decimals.

9.2 ☐ 8.8   2.19 ☐ 2.109   5.3 ☐ 4.317

3.5 ☐ 3.7   4.33 ☐ 4.37   3.07 ☐ 3.070

4.60 ☐ 4.6   8.15 ☐ 8.1   5.291 ☐ 5.298

D. Order the decimals from least to greatest. Remember to look at the whole numbers first.

6.2, 9.3, 2.9, 5.8   1.2, 4.78, 3.9, 4.75

_____   _____

3.2, 0.02, 0.576, 1.576   2.67, 47.12, 80.3, 2.495

_____   _____

0.823, 1.845, 1.8, 0.85   5.02, 2.37, 5.109, 0.506

_____   _____

Easy Peasy All-in-One Homeschool   EP Math 5/6 Workbook

LESSON 93 Practice

A. Compare the decimals.

4.3 ☐ 4.7   1.36 ☐ 1.362   9.2 ☐ 8.253

2.5 ☐ 0.7   8.02 ☐ 8.05   2.50 ☐ 2.500

5.25 ☐ 5.2   4.10 ☐ 4.1   6.426 ☐ 6.476

B. Order the decimals from least to greatest.

7.5, 8.12, 5.82, 8.13   0.73, 0.03, 0.3, 0.37

_____   _____

6.25, 12.01, 8.4, 0.113   9.2, 10.2, 5.279, 5.273

_____   _____

C. Order the decimals from greatest to least.

2.2, 2.15, 2.58, 2.5   4.87, 4.75, 4.25, 4.82

_____   _____

3.146, 4.8, 3.72, 4.827   2.37, 5.203, 2.1, 5.84

_____   _____

D. Make the smallest possible number using each set of digits. Use each digit only once. You can include a decimal point.

7, 0, 5, 1   0, 4, 0, 9   6, 0, 0, 0

_____   _____   _____

Easy Peasy All-in-One Homeschool   EP Math 5/6 Workbook

LESSON 94 Adding Decimals

A. Here are the steps for adding decimals.

```
  1 2 . 5 0 0
      4 . 0 6 3
+     9 . 7 8 0
  ─────────────
  2 6 . 3 4 3
```

To solve 12.5 + 4.063 + 9.78:
1. Line up the decimal points vertically.
2. Add zeros so the decimal parts have the same length. Add two zeros to 12.5 and one zero to 9.78.
3. Add the numbers as you would whole numbers.
4. Carry the decimal point directly down into the answer.

B. Add the decimals with tenths.

0.6 + 0.8 =   2.7 + 0.5 =

8.5 + 3.4 =   7.4 + 3.9 =

9.2 + 5.6 =   0.6 + 4.5 =

5.8 + 9.8 =   0.3 + 0.7 =

C. Add the decimals.

| 1.40 | 5.25 | 6.7 | 0.02 | 0.75 |
| + 2.38 | + 2.70 | + 4.85 | + 4.3 | + 0.95 |

| 2.718 | 0.7 | 4.253 | 5.76 | 2.478 |
| + 3.5 | + 2.459 | + 7.86 | + 8.421 | + 9.522 |

Easy Peasy All-in-One Homeschool   EP Math 5/6 Workbook

LESSON 94 Practice

A. Add the decimals with tenths.

$3.8 + 5.8 =$ $8.7 + 0.8 =$

$2.7 + 1.5 =$ $2.1 + 7.9 =$

$4.6 + 9.6 =$ $0.9 + 0.5 =$

$0.6 + 9.7 =$ $3.5 + 7.4 =$

B. Add the decimals.

```
   2.1         3.62        5.3         6.85        1.46
+  4.93     +  7.7      +  2.09     +  4.9      +  3.85
```

```
   3.807       0.5         2.598       9.57        0.798
+  7.6      +  2.509    +  7.15     +  8.643    +  0.004
```

```
   7.3         6.215       4.38        0.023       8.524
+  9.482    +  5.9      +  5.659    +  3.48     +  1.476
```

Easy Peasy All-in-One Homeschool   EP Math 5/6 Workbook

LESSON 95 Adding Decimals

A. A whole number has no decimal part, so all decimal places after the decimal point are equal to zero. Note that all of the following numbers are equal.

$$43 = 43.0 = 43.00 = 43.000 = 43.0000000$$

B. Here are the steps for adding decimals and whole numbers.

```
  1 2 . 0 0 0
+     4 . 1 6 3
  ─────────────
  2 6 . 1 6 3
```

To solve 12 + 4.163:
1. Express whole numbers as decimals by adding a decimal point and zeros. 12 is expressed as 12.000.
2. Add the decimals as usual.

C. Add the decimals.

42 + 17.3 =   42.0   80.527 + 14.5 =   80.527
              + 17.3   + 14.500

56.8 + 29 =   29.5 + 52.853 =

37 + 18.54 =   23.11 + 36.89 =

72.49 + 16 =   86.53 + 6.034 =

25 + 8.203 =   39.258 + 5.27 =

Easy Peasy All-in-One Homeschool   EP Math 5/6 Workbook

LESSON 95 Practice

Add the decimals.

$55 + 7.99 =$ $50.084 + 20.8 =$

$12.9 + 72 =$ $69.8 + 84.304 =$

$73 + 61.89 =$ $40.65 + 96.78 =$

$14.94 + 67 =$ $39.49 + 2.853 =$

$39 + 2.905 =$ $27.475 + 3.83 =$

$8.052 + 78 =$ $4.305 + 82.07 =$

$6.75 + 8.27 =$ $99.99 + 9.999 =$

LESSON 96 Adding Decimals

A. One of the most common places you find decimals is grocery receipts. The following receipts show average prices per pound of some common grocery items. Find the total amount of each receipt. How much would you have paid in 1913? What about 2013?

```
EP GROCERIES  1913
-----------------------------
BREAD                    0.06
FLOUR                    0.03
MILK 1 GALLON            0.36
CHEESE                   0.22
POTATOES                 0.02
SIRLOIN STEAK            0.24
BACON                    0.25
EGG 1 DOZEN              0.37
-----------------------------
TOTAL
-----------------------------
        THANK YOU!
```

```
EP GROCERIES  2013
-----------------------------
BREAD                    1.42
FLOUR                    0.52
MILK 1 GALLON            3.53
CHEESE                   5.83
POTATOES                 0.63
SIRLOIN STEAK            5.71
BACON                    4.41
EGG 1 DOZEN              1.93
-----------------------------
TOTAL
-----------------------------
        THANK YOU!
```

Source: U.S. Bureau of Labor Statistics

B. Add the decimals and use the sums to fill in the puzzle.

$1.65 + 1.055 =$ _____

$10.1 + 293.4 =$ _____

$30.9 + 15.02 =$ _____

$5.13 + 0.104 =$ _____

$26.2 + 33.22 =$ _____

$4.02 + 4.038 =$ _____

$52.4 + 440.4 =$ _____

Easy Peasy All-in-One Homeschool   EP Math 5/6 Workbook

LESSON 96 Practice

A. Add the money amounts. Don't forget to put the currency symbol in your answer.

```
    $2.00         $5.54         $8.02         $4.65
+   $3.47      +  $0.32      +  $0.16      +  $3.02
```

```
    Pound         Euro        Chinese Yuan   Russian Ruble

    £9.58         €5.64         ¥2.47         ₽7.63
+   £3.14      +  €8.42      +  ¥9.17      +  ₽2.20
```

B. Add the decimals.

64 + 5.38 =   21.613 + 78.6 =

52.5 + 74 =   15.4 + 71.874 =

84 + 69.13 =   32.47 + 73.98 =

22.93 + 58 =   26.68 + 5.629 =

Easy Peasy All-in-One Homeschool   EP Math 5/6 Workbook

LESSON 97 Subtracting Decimals

A. Here are the steps for subtracting decimals.

```
  1 2 . 5 0 0
-     4 . 0 6 3
  ─────────────
      8 . 4 3 7
```

To solve 12.5 − 4.063:
1. Line up the decimal points vertically.
2. Add zeros so the decimal parts have the same length.
3. Subtract the numbers as you would whole numbers.
4. Carry the decimal point directly down into the answer.

B. Subtract the decimals with tenths.

$0.9 - 0.2 =$ $\qquad$ $20.8 - 10.5 =$

$7.5 - 5.4 =$ $\qquad$ $48.4 - 23.9 =$

$9.2 - 5.6 =$ $\qquad$ $90.6 - 18.6 =$

C. Subtract the decimals.

5.30	6.26	4.3	9.22	0.92
− 2.38	− 4.70	− 0.82	− 3.8	− 0.06

4.061	7.7	8.103	0.92	6.002
− 1.5	− 5.809	− 3.86	− 0.183	− 2.586

A math riddle for you! What mathematical symbol can be placed between 4 and 5 to get a number larger than 4 and smaller than 5?

Easy Peasy All-in-One Homeschool $\qquad$ EP Math 5/6 Workbook

LESSON 97 Practice

A. Subtract the decimals with tenths.

$5.6 - 3.6 =$ $50.7 - 20.3 =$

$2.4 - 0.7 =$ $26.5 - 11.9 =$

$9.2 - 7.6 =$ $93.4 - 36.4 =$

$6.3 - 2.5 =$ $74.5 - 17.9 =$

B. Subtract the decimals.

4.1	7.72	5.2	6.38	8.42
− 3.93	− 5.8	− 1.64	− 4.9	− 0.79

7.807	9.5	7.903	9.52	0.415
− 2.4	− 8.829	− 4.39	− 8.043	− 0.018

4.3	9.005	6.23	8.023	5.532
− 2.074	− 3.5	− 5.548	− 2.49	− 5.478

Easy Peasy All-in-One Homeschool   EP Math 5/6 Workbook

LESSON 98 Subtracting Decimals

A. In real life, we commonly subtract decimals from whole numbers to make change. Determine how much change you would receive in each situation below.

You buy an item at $2.53 and pay with a $5 bill.

_____

You buy an item at $11.99 and pay with a $20 bill.

_____

You pay a check of $22.87 with three $10 bills.

_____

You pay a check of $65.40 with four $20 bills.

_____

B. Subtract the decimals.

$42 - 15.3 =$ 		42.⓪
                  − 15.3
                  _____

$50.327 - 14.5 =$ 	50.327
                  − 14.5⓪⓪
                  _____

$56.8 - 29 =$ 		$52.5 - 29.853 =$

$37 - 18.54 =$ 		$36.11 - 23.89 =$

$70.49 - 28 =$ 		$82.03 - 6.534 =$

$35 - 9.103 =$ 		$39.255 - 5.27 =$

Easy Peasy All-in-One Homeschool 		EP Math 5/6 Workbook

LESSON 98 Practice

A. Determine how much change you would receive in each situation.

You buy an item at $3.99 and pay with a $10 bill.   _____

You buy an item at $9.25 and pay with a $20 bill.   _____

You pay a check of $12.43 with three $5 bills.   _____

You pay a check of $35.90 with two $20 bills.   _____

B. Subtract the decimals.

$54 - 7.99 =$ $50.084 - 20.8 =$

$72.9 - 18 =$ $84.3 - 48.704 =$

$73 - 26.89 =$ $96.25 - 30.78 =$

$64.94 - 37 =$ $39.49 - 2.853 =$

$59 - 2.905 =$ $46.202 - 3.84 =$

LESSON 99 Word Problems: Adding/Subtracting Decimals

Solve the word problems.

Ana used 2.4 gallons of red paint and 3.9 gallons of white paint to paint her room. How much paint did she use in all? _____

Dana bought $43.12 in groceries at a store and paid with a $50 bill. How much change did she receive? _____

Mia saved $36.25 last month and $28.59 this month. How much money did she save in total? _____

David went for a run on a 4-mile trail. He ran 2.7 miles and walked the rest. How many miles did he walk? _____

A video game costs $37.10. It's on sale for $24.99. How much less is the sale price than the original price? _____

Atticus has $24.35. Matt has $19.46 more than Atticus. Naomi has $12.58 less than Matt. How much money does Naomi have? _____

This morning 1.7 inches of snow fell. By afternoon 0.8 inch of snow melted away. How much snow is left? _____

The international standard A4 paper is 8.27 inches wide and 11.69 inches long. What is the perimeter of A4 paper? _____

Fresh water freezes at 32 °F, but seawater freezes at about 28.4 °F. At how much lower of a temperature does sea water freeze? _____

Hunter weighs 32.5 kg, Max weighs 37.9 kg, and Ron weighs 29.8 kg. What is the difference between the lightest and heaviest? _____

Kirstin rode her bike 6.25 miles to the library and 8.5 miles to the park. Later she came home the same way. How many miles did Kirstin ride in all? _____

The tallest building in the world is Burj Khalifa, which is 0.828 km tall. The second tallest building is Shanghai Tower, which is 0.632 km tall. What is the difference of their heights? _____

Easy Peasy All-in-One Homeschool   EP Math 5/6 Workbook

LESSON 99 Practice

Solve the word problems.

Tom wants to buy a hat for $11.32 and a shirt for $18.79. How much money will he need to buy both?

Adam is buying a math puzzle book for $8.37. If he pays with a $10 bill, how much change will he get back?

Last night 3.4 inches of snow fell. This morning 1.8 inches of more snow fell. How much snow is on the ground?

A tank had 8 gallons of water. Kyle used 2.6 gallons of water from the tank. How much water was left in the tank?

Kate ran 2.6 miles yesterday. She ran 3.75 miles today. How many miles did she run in total?

At a store, Jessica paid her groceries with two $20 bills and received $1.46 as change. What was the cost of her groceries?

Owen bought three books at $12.65, $13.78, and $10.89. How much money did Owen spend in all?

Seth's room is 120.85 square feet. Mia's room is 140.18 square feet. How much bigger is Mia's room than Seth's?

This morning the temperature was 62.34 °F. In the evening, it was 55.90 °F. How far did the temperature drop during the day?

US Letter paper is 21.59 centimeters wide and 27.94 centimeters long. What is the perimeter of US Letter paper?

An ostrich can run up to 96.6 kilometers per hour. A horse can run 70.76 kilometers per hour. How much faster can the ostrich run than the horse?

Light from the sun reaches Earth in 8.3 minutes. It reaches Saturn in 79.3 minutes. How many more minutes does it take to reach Saturn than Earth?

LESSON 100 Multiplying by Powers of 10

A. A **power of 10** is 10 multiplied by itself a certain number of times. To calculate a power of 10, simply put as many zeros as the exponent after the 1.

$10^2 = 10 \times 10 = 100$ $10^5 = 100,000$ (5 zeros)

$10^3 = 10 \times 10 \times 10 = 1,000$ $10^7 = 10,000,000$ (7 zeros)

B. To multiply decimals by a power of 10, move the decimal point to the right as many places as the number of zeros in the power of 10. The following table shows the place value pattern when decimals are multiplied by powers of 10. Complete the table.

	42.5	3.27	0.069	0.0083
× 10 or 10^1	425		0.69	
× 100 or 10^2				0.83
× 1000 or 10^3		3,270		

C. Multiply the decimals by a power of 10.

$0.214 \times 10^2 =$ $0.002 \times 10^2 =$

$0.508 \times 10^4 =$ $3.404 \times 10^3 =$

$0.2746 \times 10^2 =$ $0.6712 \times 10^5 =$

$0.0628 \times 10^5 =$ $5.4203 \times 10^3 =$

D. Fill in the blank to make each statement true.

$0.15 \times$ ____ $= 15$ ____ $\times 10^2 = 3.8$

$0.002 \times$ ____ $= 0.02$ ____ $\times 10^3 = 973$

$72.50 \times$ ____ $= 7,250$ ____ $\times 10^5 = 9,200$

Easy Peasy All-in-One Homeschool   EP Math 5/6 Workbook

LESSON 100 Practice

A. Evaluate each expression.

$10^1 =$ $10^6 =$

$10^4 =$ $10^9 =$

B. Multiply the decimals by a power of 10.

$0.0762 \times 10 =$ $49.9 \times 10^2 =$

$0.0762 \times 10^2 =$ $5.816 \times 10^3 =$

$0.0762 \times 10^3 =$ $0.329 \times 10^4 =$

$0.0762 \times 10^4 =$ $0.0573 \times 10^2 =$

$0.0762 \times 10^5 =$ $0.0004 \times 10^7 =$

C. Fill in the blank to make each statement true.

$0.327 \times \underline{} = 3.27$ $\underline{} \times 10^2 = 63.8$

$0.002 \times \underline{} = 2{,}000$ $\underline{} \times 10^4 = 4{,}200$

$3.008 \times \underline{} = 300.8$ $\underline{} \times 10^5 = 97{,}400$

D. Scientists write numbers in a special way called **scientific notation** to make them easy to work with. In scientific notation, a number is written in the form $a \times 10^n$, where $1 \leq a < 10$ and n is an integer. Using this notation, you can shorten very large numbers in a consistent form. Let's try writing some numbers in scientific notation.

Distance from the earth to the moon: 240,000 miles = 2.4 x 10⁵

Population of the world: 7,000,000,000 = \underline{}

The sun's core temperature: 27,000,000 °F = \underline{}

Easy Peasy All-in-One Homeschool   EP Math 5/6 Workbook

LESSON 101 Multiplying Decimals

A. Here are the steps for multiplying decimals. Review Lessons 21 and 22 if you need a reminder on how to multiply multi-digit numbers.

```
      1 2
        1
   1 2 . 5
 × 0 . 5 3
   ─────────
     3 7 5   (125 × 3)
   6 2 5     (125 × 5)
   ─────────
   6 . 6 2 5
```

To solve 12.5 × 0.53:
1. Ignore the decimal point and multiply as usual.
2. Count the total number of decimal places in the factors. 12.5 and 0.53 have a total of 3 decimal places.
3. Place a decimal point in the answer so that the answer has the same number of decimal places as you counted. Place a decimal point in 6625 so that it has 3 decimal places.

B. Multiply the decimals.

```
        3 2
        1 1
        6 4
      0.675            31.7              0.65             47.6
    ×   4.28        ×  0.92           ×  0.48          ×  26.4
     ──────
       5400
      13500
     270000
     ──────
     2.88900
```

```
      7.368            5.42              0.78            0.095
    ×  42.8         ×  1.225          ×  796.3        ×  25.17
     ──────         ──────            ──────           ──────
```

Easy Peasy All-in-One Homeschool   EP Math 5/6 Workbook

LESSON 101 Practice

Multiply the decimals.

$$\begin{array}{r}0.08\\ \times\ \ 0.9\\ \hline\end{array} \qquad \begin{array}{r}9.6\\ \times\ \ 0.7\\ \hline\end{array} \qquad \begin{array}{r}0.62\\ \times\ \ \ 8.4\\ \hline\end{array} \qquad \begin{array}{r}1.79\\ \times\ \ \ 2.3\\ \hline\end{array} \qquad \begin{array}{r}20.9\\ \times\ 0.42\\ \hline\end{array}$$

$$\begin{array}{r}3.82\\ \times\ \ 47.6\\ \hline\end{array} \qquad \begin{array}{r}50.6\\ \times\ \ 0.114\\ \hline\end{array} \qquad \begin{array}{r}3.29\\ \times\ \ 4.15\\ \hline\end{array} \qquad \begin{array}{r}0.672\\ \times\ \ \ 3.07\\ \hline\end{array}$$

$$\begin{array}{r}0.3645\\ \times\ \ \ \ 12.8\\ \hline\end{array} \qquad \begin{array}{r}605.2\\ \times\ \ 73.4\\ \hline\end{array} \qquad \begin{array}{r}0.047\\ \times\ \ 27.16\\ \hline\end{array} \qquad \begin{array}{r}0.38\\ \times\ \ 440.5\\ \hline\end{array}$$

LESSON 102 Dividing by Powers of 10

A. To divide decimals by a power of 10, move the decimal point to the left as many places as the number of zeros in the power of 10. The following table shows the place value pattern when decimals are divided by powers of 10. Remember that whole numbers have an implied decimal point at the end. Complete the table.

	8,500	463	10.3	2.6
$\div 10$ or 10^1	850		1.03	
$\div 100$ or 10^2				0.026
$\div 1000$ or 10^3		0.463		

B. Divide the decimals by a power of 10.

$703 \div 10^2 =$ $\qquad$ $2.5 \div 10^2 =$

$22.6 \div 10^3 =$ $\qquad$ $340 \div 10^4 =$

$574.5 \div 10^2 =$ $\qquad$ $6,710 \div 10^5 =$

$12,800 \div 10^5 =$ $\qquad$ $14,203 \div 10^3 =$

C. Fill in the blank to make each statement true.

$0.42 \div \underline{\qquad} = 0.042$ $\qquad$ $\underline{\qquad} \div 10^2 = 2.47$

$25.6 \div \underline{\qquad} = 0.256$ $\qquad$ $\underline{\qquad} \div 10^2 = 0.988$

$7,250 \div \underline{\qquad} = 0.725$ $\qquad$ $\underline{\qquad} \div 10^5 = 0.0306$

D. Solve the word problems.

A roll of 100 stamps costs $49. What's the price of one stamp? $\underline{\qquad}$

Randy bought 10 stamps with $20. What was his change? $\underline{\qquad}$

Easy Peasy All-in-One Homeschool   EP Math 5/6 Workbook

LESSON 102 Practice

A. Divide the decimals by a power of 10.

$86{,}200 \div 10 =$ $\qquad$ $16 \div 10^4 =$

$86{,}200 \div 10^2 =$ $\qquad$ $4.2 \div 10^2 =$

$86{,}200 \div 10^3 =$ $\qquad$ $273 \div 10^3 =$

$86{,}200 \div 10^4 =$ $\qquad$ $5{,}800 \div 10^5 =$

$86{,}200 \div 10^5 =$ $\qquad$ $78{,}253 \div 10^3 =$

B. Fill in the blank to make each statement true.

$0.64 \div \underline{\quad} = 0.064$ $\qquad$ $\underline{\quad} \div 10^2 = 3.07$

$50.2 \div \underline{\quad} = 0.502$ $\qquad$ $\underline{\quad} \div 10^2 = 0.158$

$62.8 \div \underline{\quad} = 0.0628$ $\qquad$ $\underline{\quad} \div 10^3 = 0.425$

$4{,}300 \div \underline{\quad} = 0.0043$ $\qquad$ $\underline{\quad} \div 10 = 1.203$

$7{,}029 \div \underline{\quad} = 0.7029$ $\qquad$ $\underline{\quad} \div 10^4 = 0.764$

C. Solve the word problems.

A local restaurant bought 10 pounds of beef at $53.20. What is the price of beef per pound?   _____

A box of 100 pencils sells at $12.00. Jason bought 25 pencils and paid $10. What was his change?   _____

A supermarket sells 20 pounds of potatoes at $38.60. How much does 2 pounds of potatoes cost?   _____

Joan's car can drive 100 miles on 6 gallons of gas. How many gallons of gas does it use per mile?   _____

Easy Peasy All-in-One Homeschool   EP Math 5/6 Workbook

LESSON 103 Dividing Decimals

A. Here are the steps for dividing decimals.

$$0.06 \overline{)0.468} \Rightarrow 6 \overline{)46.8}$$

```
      7.8
  6 ) 46.8
      42
      ---
       48
       48
       ---
        0
```

Move both decimal points 2 places to the right.

To solve $0.468 \div 0.06$:
1. Move the decimal points in the divisor and dividend the same number of places to the right to make the divisor a whole number.
2. Ignore the decimal points and divide as usual.
3. Place a decimal point in the quotient directly above the decimal point in the dividend.

B. Divide the decimals.

$0.9 \overline{)6.3}$ quotient 7; 63, 0

$0.04 \overline{)23.60}$ quotient 590; 20, 36, 36, 00

$0.7 \overline{)5.81}$

$0.62 \overline{)8.06}$

$0.5 \overline{)4.5}$

$0.6 \overline{)4.98}$

$0.26 \overline{)9.62}$

$0.13 \overline{)71.5}$

$0.8 \overline{)6.4}$

$1.2 \overline{)3.12}$

$0.42 \overline{)75.6}$

$0.06 \overline{)5.04}$

Easy Peasy All-in-One Homeschool   EP Math 5/6 Workbook

LESSON 103 Practice

A. Divide the decimals.

0.3) 2.7   0.6) 1.44   0.07) 12.6   0.06) 2.88

0.6) 3.6   0.4) 3.72   0.23) 3.22   0.53) 95.4

0.7) 2.8   1.8) 2.52   0.48) 62.4   0.12) 5.76

B. Solve the word problems. (Hint: The area of a rectangle is its width times its length.)

Carol is making a rectangular game board. Its area is 388.8 square inches, and one side is 18 inches long. How long is the other side? _____

Carol wants to cover the board with 3.6-inch square tiles. How many tiles will she need? _____

Carol decided to put a ribbon stripe along the perimeter of the board. How many inches of ribbon will she use? _____

LESSON 104 Dividing Decimals

A. Here's an example of dividing decimals where there is a remainder.

$$0.6 \overline{\smash{)}0.9} \Rightarrow 6 \overline{\smash{)}9.0}$$

Move both decimal points 1 place to the right.

```
      1.5
   6 ) 9.0
       6
       ――
       3 0
       3 0
       ――
         0
```

To solve 0.9 ÷ 0.6:
1. Move the decimal points as before.
2. Ignore the decimal points and divide as usual.
3. If there is a remainder, keep adding zeros to the right of the dividend and continue to divide.
4. Place a decimal point in the quotient where appropriate.

B. Divide the decimals.

$0.03 \overline{\smash{)}8.067}$ $0.45 \overline{\smash{)}5.193}$ $0.68 \overline{\smash{)}55.794}$

$0.05 \overline{\smash{)}1.743}$ $0.18 \overline{\smash{)}7.965}$ $0.42 \overline{\smash{)}85.134}$

Easy Peasy All-in-One Homeschool   EP Math 5/6 Workbook

LESSON 104 Practice

Divide the decimals.

0.8 | 1.2   0.7 | 6.02   0.34 | 40.8   0.61 | 1.22

0.6 | 2.7   3.2 | 2.56   0.45 | 23.4   0.08 | 7.68

0.32 | 2.576   0.54 | 68.04   0.77 | 429.66

0.55 | 15.18   0.42 | 3.885   0.31 | 15.748

LESSON 105 Converting Fractions to Decimals

A. A decimal is a special way of writing a fraction whose denominator is a power of 10. Here are some examples. Fill in the numerators.

$$0.7 = \frac{7}{10} \qquad 0.53 = \frac{53}{100} \qquad 0.029 = \frac{}{1000}$$

$$2.3 = \frac{23}{10} \qquad 4.09 = \frac{409}{100} \qquad 5.103 = \frac{}{1000}$$

B. One way to convert a fraction to a decimal is to find an equivalent fraction whose denominator is a power of 10. Then write the equivalent fraction as a decimal. Complete the second example.

$$\frac{3}{5} = \frac{3 \times 2}{5 \times 2} = \frac{6}{10} = 0.6 \qquad \frac{31}{25} = \frac{31 \times 4}{25 \times 4} = \frac{}{100} =$$

C. It is always a good practice to reduce fractions before performing any operation.

$$\frac{28}{80} = \frac{28 \div 4}{80 \div 4} = \frac{7}{20} = \frac{7 \times 5}{20 \times 5} = \frac{35}{100} = 0.35$$

D. Convert the fractions to decimals. Remember to reduce fractions when possible!

$$\frac{1}{2} = \qquad\qquad \frac{51}{50} =$$

$$\frac{7}{5} = \qquad\qquad \frac{63}{75} =$$

$$\frac{9}{25} = \qquad\qquad \frac{58}{200} =$$

$$\frac{27}{20} = \qquad\qquad \frac{136}{250} =$$

Easy Peasy All-in-One Homeschool   EP Math 5/6 Workbook

LESSON 105 Practice

A. Write each decimal as a fraction with a power of 10 in the denominator.

0.9 =   0.07 =   0.217 =

6.5 =   3.29 =   1.403 =

B. Convert the fractions to decimals.

$\dfrac{5}{2} =$ $\dfrac{39}{75} =$

$\dfrac{7}{5} =$ $\dfrac{45}{50} =$

$\dfrac{6}{4} =$ $\dfrac{11}{20} =$

$\dfrac{24}{15} =$ $\dfrac{53}{25} =$

$\dfrac{16}{80} =$ $\dfrac{72}{200} =$

$\dfrac{37}{50} =$ $\dfrac{96}{300} =$

Did you know? The name Google came from misspelling the word 'googol', which is a very large number. A googol is the number 1 followed by 100 zeros.

Easy Peasy All-in-One Homeschool   EP Math 5/6 Workbook

LESSON 106 Converting Fractions to Decimals

A. A **fraction** is a division expression where the numerator is divided by the denominator. To convert any fraction to a decimal, simply divide the numerator by the denominator. On the right is an example of using long division to convert fractions to decimals.

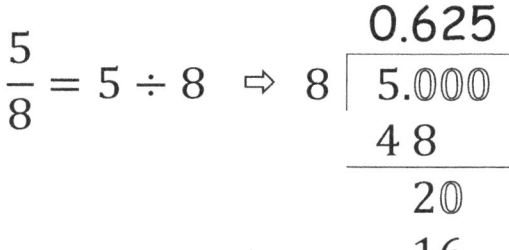

$$\frac{5}{8} = 5 \div 8 \Rightarrow 8\overline{)5.000} = 0.625$$

B. Convert the fractions to decimals. Round off your answers to 3 decimal places. (Rounding off decimals to 3 decimal places means rounding to the nearest thousandth.)

$\dfrac{3}{8} =$

$\dfrac{38}{4} =$

$\dfrac{14}{25} =$

$\dfrac{49}{35} =$

$\dfrac{3}{7} =$

$\dfrac{14}{3} =$

$\dfrac{10}{11} =$

$\dfrac{42}{16} =$

Easy Peasy All-in-One Homeschool   EP Math 5/6 Workbook

LESSON 106 Practice

Convert the fractions to decimals. Round off decimals to 3 decimal places.

$\dfrac{9}{4} =$ $\qquad$ $\dfrac{33}{6} =$ $\qquad$ $\dfrac{53}{20} =$ $\qquad$ $\dfrac{44}{80} =$

$\dfrac{7}{8} =$ $\qquad$ $\dfrac{41}{5} =$ $\qquad$ $\dfrac{15}{40} =$ $\qquad$ $\dfrac{30}{48} =$

$\dfrac{5}{9} =$ $\qquad$ $\dfrac{20}{7} =$ $\qquad$ $\dfrac{25}{15} =$ $\qquad$ $\dfrac{33}{28} =$

Easy Peasy All-in-One Homeschool · EP Math 5/6 Workbook

LESSON 107 Converting Decimals to Fractions

A. Here are the steps for converting decimals to fractions. Complete the examples.

$$0.25 = \frac{25}{100} = \frac{\ }{4}$$

$$7.048 = 7\frac{48}{1000} = 7\frac{\ }{125}$$

1. Leave the whole number part as is.
2. Convert the decimal part to a fraction where the numerator is the number after the decimal point and the denominator is a power of 10 with as many zeros as the number of decimal places.
3. Simplify the fraction.

B. Convert the decimals to fractions. Simplify your answers.

$0.6 =$ $\qquad\qquad\qquad\qquad$ $1.22 =$

$9.8 =$ $\qquad\qquad\qquad\qquad$ $8.45 =$

$0.05 =$ $\qquad\qquad\qquad\qquad$ $0.076 =$

$0.28 =$ $\qquad\qquad\qquad\qquad$ $0.275 =$

$2.75 =$ $\qquad\qquad\qquad\qquad$ $3.625 =$

$1.36 =$ $\qquad\qquad\qquad\qquad$ $5.008 =$

Easy Peasy All-in-One Homeschool $\qquad\qquad\qquad\qquad$ EP Math 5/6 Workbook

LESSON 107 Practice

Convert the decimals to fractions. Simplify your answers.

0.2 =   8.02 =

4.9 =   0.82 =

0.8 =   5.66 =

7.6 =   0.148 =

0.24 =   3.005 =

0.95 =   0.872 =

1.27 =   4.635 =

6.75 =   9.028 =

LESSON 108 Converting Decimals to Fractions

A. Converting decimals to fractions is simply a matter of using appropriate denominators. Complete the examples.

$0.4 = \dfrac{4}{10} = \dfrac{2}{5}$ $\qquad$ $10.56 = 10\dfrac{56}{100} = 10\dfrac{14}{}$

$3.5 = 3\dfrac{5}{10} = 3\dfrac{1}{}$ $\qquad$ $0.072 = \dfrac{72}{1000} = \dfrac{9}{125}$

B. Convert the decimals to fractions. Simplify your answers.

$0.2 =$ $\qquad\qquad\qquad$ $1.44 =$

$3.1 =$ $\qquad\qquad\qquad$ $7.94 =$

$0.04 =$ $\qquad\qquad\qquad$ $0.002 =$

$0.48 =$ $\qquad\qquad\qquad$ $0.035 =$

$4.25 =$ $\qquad\qquad\qquad$ $1.875 =$

$6.15 =$ $\qquad\qquad\qquad$ $4.025 =$

Easy Peasy All-in-One Homeschool $\qquad\qquad$ EP Math 5/6 Workbook

LESSON 108 Practice

Convert the decimals to fractions. Simplify your answers.

0.3 =   4.05 =

2.8 =   0.54 =

3.2 =   1.88 =

4.3 =   0.125 =

0.22 =   7.008 =

0.65 =   0.325 =

1.49 =   9.504 =

7.75 =   8.027 =

LESSON 109 Converting Fractions to Repeating Decimals

A. Some fractions are converted to decimals where a digit or a group of digits repeats without ending. Such decimals are called **repeating decimals**, and are represented by putting a horizontal bar above the repeating digit(s). Here are some examples.

$$\frac{1}{3} = 0.3333... = 0.\overline{3} \qquad \frac{2}{11} = 0.181818... = 0.\overline{18}$$

$$\frac{1}{6} = 0.1666... = 0.1\overline{6} \qquad \frac{1}{7} = 0.\overline{142857}$$

B. Convert the fractions to decimal. If necessary, use a bar to indicate the repeating digit(s).

$$\frac{2}{3} = \qquad \frac{1}{9} = \qquad \frac{14}{16} = \qquad \frac{50}{66} =$$

$$\frac{5}{6} = \qquad \frac{7}{11} = \qquad \frac{29}{37} = \qquad \frac{54}{80} =$$

Easy Peasy All-in-One Homeschool — EP Math 5/6 Workbook

LESSON 109 Practice

Convert the fractions to decimal. If necessary, use a bar to indicate the repeating digit(s).

$\dfrac{4}{3} =$ $\dfrac{1}{8} =$ $\dfrac{1}{11} =$ $\dfrac{4}{33} =$

$\dfrac{7}{6} =$ $\dfrac{4}{18} =$ $\dfrac{9}{12} =$ $\dfrac{27}{20} =$

$\dfrac{9}{4} =$ $\dfrac{8}{12} =$ $\dfrac{41}{90} =$ $\dfrac{31}{99} =$

Easy Peasy All-in-One Homeschool   EP Math 5/6 Workbook

LESSON 110 The Meaning of Percent

A. Percent means "per 100" or "out of 100" and we use the symbol % to represent it. A percent is a special way of writing a fraction with a denominator of 100. Here are some examples of expressing the shaded portion of each grid as a fraction, decimal, and percent.

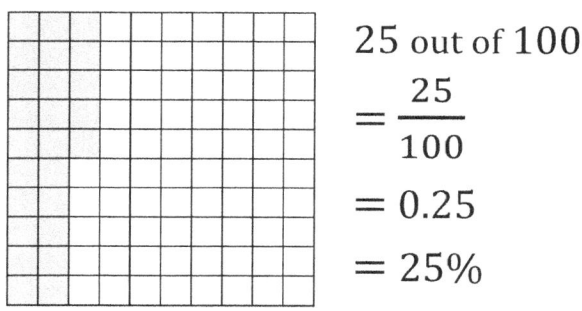

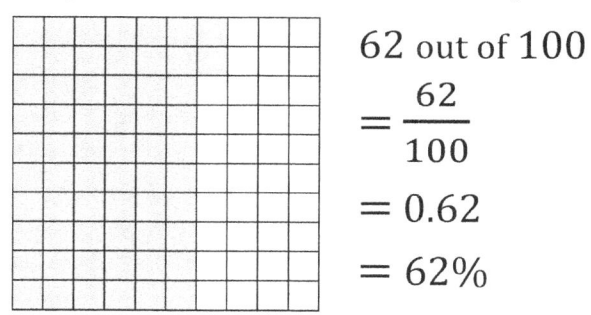

25 out of 100
$= \dfrac{25}{100}$
$= 0.25$
$= 25\%$

62 out of 100
$= \dfrac{62}{100}$
$= 0.62$
$= 62\%$

B. Fractions, decimals, and percents are all used to represent parts of a whole. They are just different ways of writing the same quantity, as shown in the examples below.

4 out of 10 $= \dfrac{4}{10} = \dfrac{40}{100} = 0.40 = 40\%$

125 out of 1000 $= \dfrac{125}{1000} = \dfrac{12.5}{100} = 0.125 = 12.5\%$

C. Express each quantity as a fraction, decimal, and percent.

	Fraction	Decimal	Percent
5 out of 10			
8 out of 10			
36 out of 100			
25 out of 1000			
860 out of 1000			

Easy Peasy All-in-One Homeschool   EP Math 5/6 Workbook

LESSON 110 Practice

Express each quantity as a fraction, decimal, and percent.

	Fraction	Decimal	Percent
1 out of 10			
7 out of 10			
6 out of 10			
5 out of 100			
75 out of 100			
90 out of 100			
8 out of 1000			
45 out of 1000			
192 out of 1000			
560 out of 1000			

Did you know? 111111111 x 111111111 = 12345678987654321

Easy Peasy All-in-One Homeschool · EP Math 5/6 Workbook

LESSON 111 Converting Decimals to Percents

A. To convert a decimal to a percent, multiply the decimal by 100 and add the % sign. This is equivalent to moving the decimal point 2 places to the right. Convert the decimals below.

0.5 = 50%	0.01 = 1%	0.002 = 0.2%
0.8 =	0.03 =	0.005 =
0.2 =	0.98 =	0.074 =
2.2 =	4.12 =	3.102 =

B. Express each quantity as a decimal and percent.

	Decimal	Percent	Work area
7 out of 10			
9 out of 20			
26 out of 50			
8 out of 100			
14 out of 200			
55 out of 500			
752 out of 1000			

LESSON 111 Practice

A. Convert the decimals to percents.

0.9 =	0.05 =	0.003 =
0.1 =	0.62 =	0.016 =
1.6 =	3.02 =	5.208 =
3.7 =	7.15 =	6.734 =
0.5 =	2.36 =	0.008 =
4.2 =	0.01 =	1.059 =

B. Express each quantity as a decimal and percent.

	Decimal	Percent	Work area
1 out of 10			
7 out of 20			
5 out of 100			
12 out of 500			
9 out of 1000			
274 out of 1000			

Easy Peasy All-in-One Homeschool   EP Math 5/6 Workbook

LESSON 112 Converting Percents to Decimals

A. To convert a percent to a decimal, divide the percent by 100 and drop the % sign. This is equivalent to moving the decimal point 2 places to the left. Convert the percents below.

3% = 0.03   10% = 0.1   0.3% = 0.003

1% =   40% =   7.5% =

5% =   28% =   200% =

9% =   59% =   54.7% =

B. To find a percent of a number, convert the percent to either a fraction or a decimal. Then multiply the fraction or decimal by the number. The following examples show why you can use either fractions or decimals. Complete the second example.

$$50\% \text{ of } 20 = \frac{50}{100} \text{ of } 20 = \frac{50}{100} \times 20 = 0.5 \times 20 = 10$$

$$40\% \text{ of } 45 = \frac{40}{100} \text{ of } 45 =$$

C. Calculate the percent of each number.

20% of 5   7% of 150

9% of 50   50% of 80

10% of 20   55% of 60

Easy Peasy All-in-One Homeschool   EP Math 5/6 Workbook

LESSON 112 Practice

A. Convert the percents to decimals.

2% =   11% =   0.6% =

7% =   55% =   7.2% =

3% =   32% =   100% =

5% =   29% =   240% =

B. Calculate the percent of each number.

3% of 70   6% of 220

5% of 65   5% of 180

10% of 50   4% of 675

11% of 15   22% of 90

20% of 90   70% of 40

LESSON 113 Fractions and Decimals on a Number Line

A. We can now place any fractions and decimals on a number line. Negative decimals are placed in the same way that negative fractions are placed. Here are some fractions and decimals placed on number lines. Fill in the missing numbers.

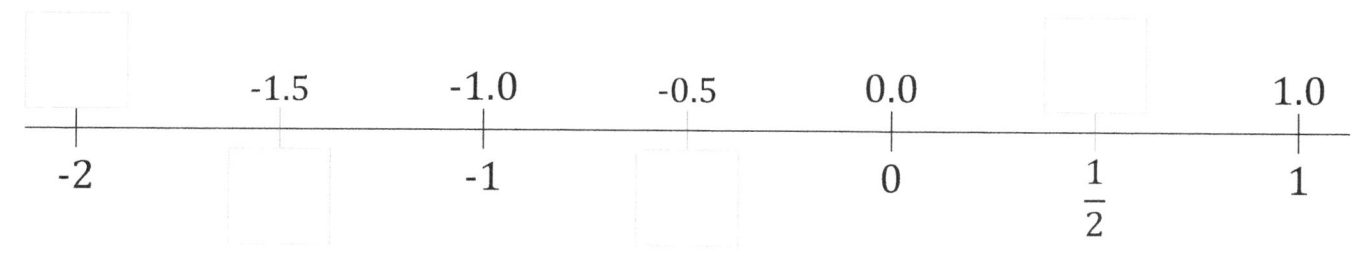

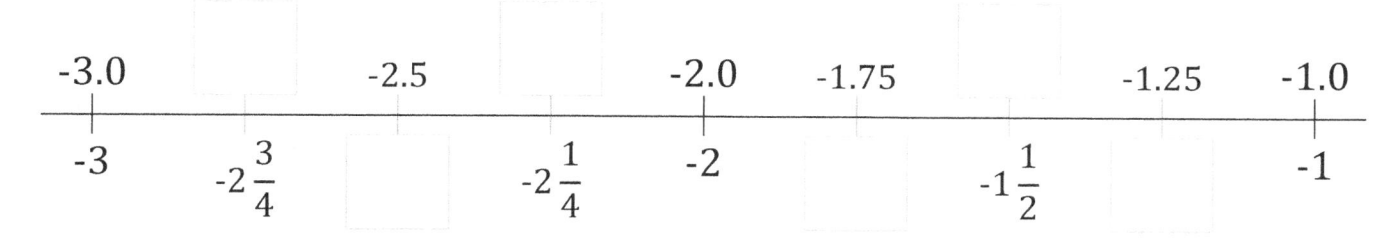

B. Match the location of each letter with the correct fraction.

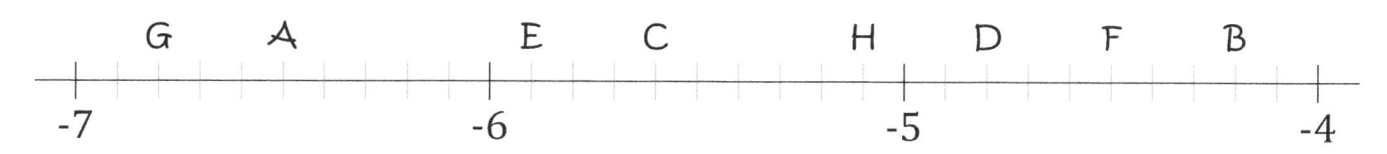

-4.8   -5.1   -4.2   -5.9

$-5\dfrac{3}{5}$ $-6\dfrac{4}{5}$ $-4\dfrac{1}{2}$ $-6\dfrac{1}{2}$

C. Write the location of each letter as a mixed number and a decimal. Simplify fractions.

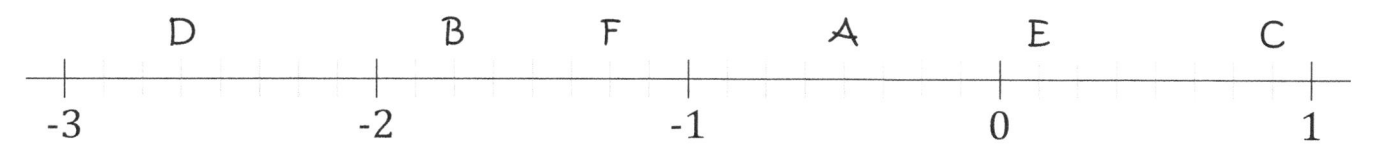

A

B

C

D

E

F

Easy Peasy All-in-One Homeschool   EP Math 5/6 Workbook

LESSON 113 Practice

A. Choose a reasonable value for the letter P on each number line.

-1.5 — P — 0.0	$-\dfrac{1}{4}$ -1.0 -0.5 $-\dfrac{8}{5}$
-5.0 — P — 1.0	-1.0 -4.0 -2.0 -0.5
-8.2 — P — -7.9	-8.0 $-8\dfrac{1}{4}$ -7.8 $-7\dfrac{2}{5}$
-0.4 — P — -0.2	-0.33 -0.46 -0.36 -0.23

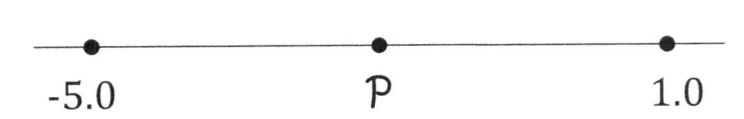

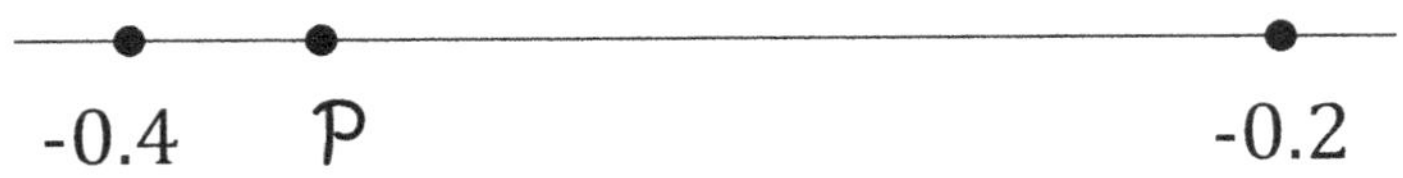

B. Write the location of each letter as a mixed number and a decimal. Simplify fractions.

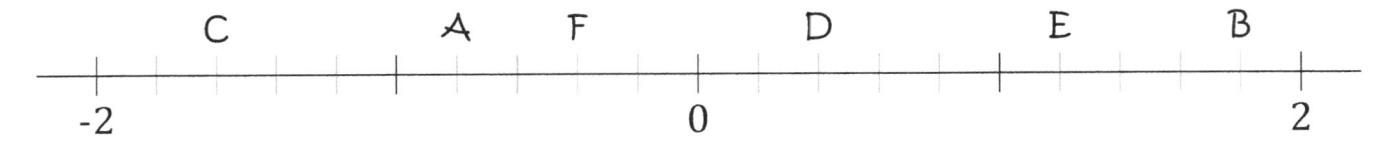

Number line from -2 to 2 with points: C, A, F, D, E, B

A   B   C

D   E   F

Number line from -10 to -6 with points: M, P, Q, N, O, L

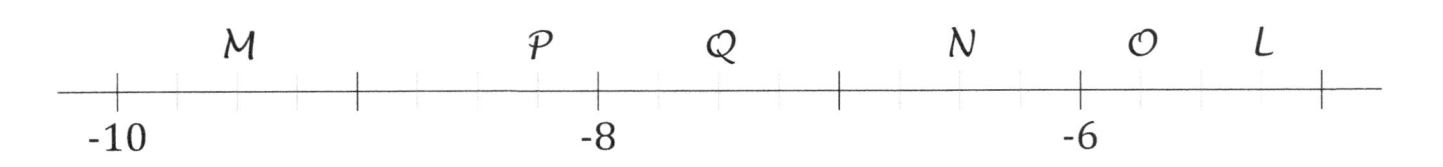

L   M   N

O   P   Q

Easy Peasy All-in-One Homeschool   EP Math 5/6 Workbook

Lesson 114 Ordering Numeric Expressions

A. You can compare numbers expressed as fractions, decimals, and percents since they are just different ways of writing the same quantity. To compare these different numeric expressions, first convert them to the same form and then compare their values. Here are some examples of using this method. Complete the examples using <, >, or =.

$\dfrac{1}{8}$ (?) 0.12 ⇨ 0.125 ( ) 0.12   or $\dfrac{1}{8}$ ( ) $\dfrac{12}{100}$

5.4% (?) 0.54 ⇨ 0.054 ( ) 0.54   or   5.4% ( ) 54%

B. Order each set of numbers from least to greatest.

0.85, 2.3, 50%   0.4, 23%, 16%

9%, 1.3, 72%   4.8, 30%, 2.5

$\dfrac{3}{5}$, 0.4, 20%   0.8, $\dfrac{5}{4}$, 25%

$\dfrac{3}{8}$, 0.125, 62% $\dfrac{11}{20}$, $\dfrac{3}{2}$, 50%

Easy Peasy All-in-One Homeschool   EP Math 5/6 Workbook

LESSON 114 Practice

Order each set of numbers from least to greatest.

1.5, 0.5, 15%   7%, 0.02, 0.05

2.4, 13%, 0.17   0.4, 32%, 28%

1.8, 2.1, 90%   2.4, 20%, 1.5

$\frac{9}{4}$, 2.3, 15%   1.2, $\frac{9}{10}$, 85%

0.72, $\frac{71}{10}$, 7% $\frac{9}{20}$, $\frac{13}{10}$, 50%

$\frac{1}{2}$, $\frac{2}{5}$, 10% $\frac{7}{25}$, 65%, 30%

LESSON 115 Numbers as Fractions, Decimals, and Percents

A. Improper fractions and mixed numbers are converted to decimals greater than 1 and percents greater than 100%, as shown in the examples below.

$$\frac{6}{5} = 1\frac{1}{5} = 1.2 = 120\% \qquad \frac{19}{8} = 2\frac{3}{8} = 2.375 = 237.5\%$$

B. Convert between fractions, decimals, and percents. Simplify fractions and round off decimals to 3 decimal places, if necessary.

Fraction	Decimal	Percent	Work area
$\frac{5}{6}$			
		68%	
	3.5		
		87.5%	
$\frac{11}{9}$			
	0.225		
$7\frac{3}{4}$			
		49.2%	

Easy Peasy All-in-One Homeschool   EP Math 5/6 Workbook

LESSON 115 Practice

Convert between fractions, decimals, and percents. Simplify fractions and round off decimals to 3 decimal places, if necessary.

Fraction	Decimal	Percent	Work area
$\frac{2}{3}$			
		75%	
$\frac{4}{9}$			
	0.82		
	1.64		
$\frac{11}{12}$			
		38%	
$\frac{16}{7}$			
	8.26		
		4.5%	

LESSON 116 Identifying Percent, Whole, and Part

A. Percent problems have three components: **percent**, **whole** (total amount), and **part** (partial amount). The part is what you get when you take the percent of the whole. This relationship can be expressed in three formulas known as **percent formulas**.

$$part = percent \times whole \qquad percent = \frac{part}{whole} \qquad whole = \frac{part}{percent}$$

B. The first step to solve a percent problem is always to identify the three components. Here are some real-world examples. Identify the percent, whole, and part in each situation. Use a question mark to represent the unknown component.

	Percent	Whole	Part
Mia has 20 crayons, and 15% of them are broken. How many crayons are broken?	15%	20	?
Danny got 48 questions correct out of 50 in a test. What percent was that?	?	50	48
Ezra spent 50% of his money to buy a book at $15. How much money did Ezra have at first?			
Terry bought a box of 60 apples, and 15 of them were bad. What percent was that?			
Josh bought a shirt at 12% off the regular price of $24. How much did he save?			
Kelly read 4 books last month. It was 80% of what Eli read. How many books did Eli read?			

C. Decide which of the three percent formulas you need to use to find the unknown value in each situation above. Substitute the known values into the appropriate formula and write the resulting formula. Do not solve.

Use $part = percent \times whole$:

Mia's crayons: $? = 15\% \times 20$

Sam's money: $? =$

Josh's saving: $? =$

Use $percent = part/whole$:

Danny's test: $? = 48/50$

Terry's apples: $? =$

Eli's books: $? =$

LESSON 116 Practice

A. Identify the percent, whole, and part in each situation. Use a question mark to represent the unknown component.

	Percent	Whole	Part
Six percent of 50 apples in a basket are bad. How many apples are bad?			
Fifteen percent of 40 crayons in a box are broken. How many crayons are broken?			
In a soccer club, 22 of 40 kids are boys. What percentage are boys?			
Eloise got 24 questions correct out of 25 in a test. What percent was that?			
Eighty percent of 150 kids attended the baseball game. How many kids were at the game?			
The dinner bill was $40, and Mr. Kim left $7 as a tip. What percent was the tip?			
Selma saved $800 for a trip and actually spent $600. What percent did he spend?			
Arlo's new laptop was $960 plus 7% sales tax. How much sales tax did he pay?			

B. Use the appropriate percent formula to find the unknown value in each situation above. Substitute the known values into the formula and write the resulting formula. Do not solve.

Apple basket: ? =   Crayon box: ? =

Soccer club: ? =   Eloise's test: ? =

Baseball game: ? =   Kim's dinner: ? =

Selma's trip: ? =   Arlo's laptop: ? =

LESSON 117 Types of Percent Problems

A. There are three types of percent problems depending on which of the three components is missing. Regardless of the type, to solve percent problems, you first identify the percent, whole, and part. Then use one of the three percent formulas to find the unknown. Let's look at each type of problem and how to solve it.

Find the part:

What is 20% of 80?

Percent = 20%
Whole = 80
Part = Unknown

Part = Percent × Whole
= 20% × 80
= 0.2 × 80 = 16

Find the percent:

36 is what percent of 90?

Percent = Unknown
Whole = 90
Part = 36

Percent = Part/Whole
= 36/90
= 0.4 = 40%

Find the whole:

6 is 30% of what number?

Percent = 30%
Whole = Unknown
Part = 6

Whole = Part/Percent
= 6/30%
= 6/0.3 = 20

B. As shown in the third example above, to find the whole in a percent problem, you need to divide the part by the decimal form of the percent. How do you divide by a decimal? You can always use long division, or you can use fractions as shown below. Use whichever method you find easier. Review Lessons 85, 100, and 103 if needed.

$$\frac{6}{0.3} = \frac{6 \times 10}{0.3 \times 10} = \frac{60}{3} = 20 \quad \text{OR} \quad 6 \div 0.3 = 6 \div \frac{3}{10} = \overset{2}{\cancel{6}} \times \frac{10}{\cancel{3}} = 20$$

C. Solve the percent problems.

What is 40% of 900?

15 is what percent of 120?

19 is 20% of what number?

Find 16% of 240.

16 is what percent of 64?

90% of what number is 135?

Easy Peasy All-in-One Homeschool

EP Math 5/6 Workbook

Lesson 117 Practice

Solve the percent problems.

Find 75% of 300.

33 is 30% of what number?

12 is what percent of 60?

What is 50% of 112?

70% of what number is 147?

49 is what percent of 350?

What 5.5% of 1400?

35 is what percent of 125?

45 is what percent of 500?

44 is 25% of what number?

Easy Peasy All-in-One Homeschool

EP Math 5/6 Workbook

LESSON 118 Word Problems: Percents

A. Copy the three percent formulas from Lesson 116 to your notebook. Use the formulas to work out the problems in this lesson.

B. Let's look at some real-world percent problems. Study each example carefully!

A TV originally cost $350. If its price increases by 10%, what is the new price?

First, identify the percent, whole, and part:
  Whole = Original price = $350
  Percent = Percent increase = 10% = 0.1
  Part = Price increase = Unknown

Second, find the unknown:
  Part = Percent × Whole
  Price increase = 0.1 × 350 = $35

Finally, find what is being asked:
  New price
  = Original price + Price increase
  = $350 + $35 = $385

So the new price is $385.

A shirt costs $16, which is 20% off its regular price. What is the regular price?

Notice that 20% is the discount percent, which means $16 is 80% of the regular price.

First, identify the percent, whole, and part:
  Percent = 100% − 20% = 80%
  Whole = Regular price = Unknown
  Part = Sale price = $16

Second, find the unknown:
  Whole = Part/Percent
  Regular price = $16/0.8 = $20

Finally, find what is being asked:
  We have already found it!

So the regular price is $20.

C. Solve the word problems.

$8.00 candy bars are now on sale for 30% off. What will be the candy bars price now? (Hint: 30% of $8 is the discount amount.) _____

The dinner bill was $42, and Emma left a 20% tip. How much did Emma pay for the dinner? (Hint: She paid $42 + 20% of $42.) _____

A video game costs $50. It's on sale for $44. What percent of the price is discounted? (Hint: $50 − $44 is discounted from $50.) _____

Cammy bought a jacket at $11. The price was 80% off its regular price. What was the regular price? (Hint: She paid 20% of the regular price. So, $11 is 20% of what number?) _____

Max's new laptop was $980 plus 7% tax. How much did Max pay for his laptop? (Hint: He paid 7% more.) _____

Easy Peasy All-in-One Homeschool   EP Math 5/6 Workbook

LESSON 118 Practice

Solve the word problems.

Lucy bought a pair of shoes at 10% off the regular price of $45. How much did she pay? _____

The dinner bill was $58, and Daniel left a 15% tip. How much did Daniel pay for the dinner? _____

Sophia bought a computer game at 12% off the regular price of $35. How much did she save? _____

Kelly read 40% of a novel, and she now has 108 pages left. How many pages does the novel have? (Hint: She has 60% of the novel left. So, 108 is 60% of what number?) _____

There are 25 children in Josh's book club. Twelve of them wear glasses. What percent of the book club wear glasses? _____

Stacey bought a table at 35% off the regular price. She paid $81.90 for the table. What was the regular price? _____

Walter's rent was $520 per month. Starting this month, it went up by 4%. What will be his rent payment now? _____

A shoe store has a sale on sneakers for 40% off. If the sale price of a pair is $39, what is the original price? _____

Laura solved 25 decimal problems and got 23 problems correct. What percent of the problems did she get correct? _____

A pair of socks sells for $12. The sales tax is 6%. Tylor bought 5 pairs of socks. How much did he pay in total? _____

The population of Daniel's city was 50,000 two years ago. The population grew by 4% last year and then dropped by 4,000 this year. What is the population of Daniel's city this year? _____

Our homeschool group includes 25 high school students, 32 middle school students, and 23 elementary students, and 55% of the students are girls. How many boys are in our group? _____

Easy Peasy All-in-One Homeschool   EP Math 5/6 Workbook

LESSON 119 Introduction to Ratios

A. A **ratio** is a comparison of two numbers. A ratio of two numbers a and b can be written in three ways: a:b, a to b, and a/b. The ratio a:b means that for every a units of one quantity there are b units of another quantity. Write each ratio below in three ways.

	Using a colon	In words	As a fraction
There are 5 apples and 8 bananas. What is the ratio of apples to bananas?	5:8		$\dfrac{5}{8}$
There are 7 girls and 9 boys at the park. What is the ratio of boys to girls?		9 to 7	
Nic has $8. Laura has $11. What is the ratio of Laura's money to Nic's money?			
William has 6 blue pens and 7 red pens. What is the ratio of blue pens to all pens?			
In the words "Easy Peasy," what is the ratio of vowels to consonants? (Y is a vowel here!)			

B. Write each ratio using a colon and as a fraction.

What is the ratio of the number of nickels in a dollar to the number of quarters in a dollar? _____ _____

Jessica has 15 pennies, 11 nickels, 7 dimes, and 14 quarters. What is the ratio of pennies to dimes? _____ _____

What is the ratio of the number of sides in a pentagon to the number of sides in a hexagon? _____ _____

What is the ratio of the number of seconds in a minute to the number of hours in a day? _____ _____

A rectangle is 5 inches wide and 8 inches long. What is the ratio of the perimeter to the area of the rectangle? _____ _____

Terry bought a box of 40 apples, and 15 of them were bad. What is the ratio of bad apples to good apples? _____ _____

Easy Peasy All-in-One Homeschool   EP Math 5/6 Workbook

LESSON 119 Practice

A. Write each ratio in three different ways.

	Using a colon	In words	As a fraction

There are 9 peaches and 12 bananas. What is the ratio of peaches to bananas?

Mia has 16 blue pens and 7 red pens. What is the ratio of red pens to blue pens?

There are 14 girls and 25 boys at the park. What is the ratio of girls to boys?

Eli saved $22. Gavin saved $30. What is the ratio of Gavin's savings to Eli's savings?

A farm has 6 cows, 7 pigs, and 25 chickens. What is the ratio of pigs to all animals?

B. Write each ratio using a colon and as a fraction.

Mike solved 17 fraction problems and 15 decimal problems. What is the ratio of decimal problems to all problems? _____ _____

What is the ratio of the number of faces in a cube to the number of faces in a pyramid? _____ _____

A rectangle is 8 inches wide and 4 inches long. What is the ratio of the perimeter to the area of the rectangle? _____ _____

Kate has 12 pennies, 7 nickels, 18 dimes, and 3 quarters. What is the ratio of dimes to all coins? _____ _____

Penny used 5 cups of flour, 4 tablespoons of butter, 2 teaspoons of cinnamon, and 3 cups of sugar to bake cookies. What is the ratio of sugar to flour? _____ _____

Paul bought a pair of socks at $15, a hat at $13, and a shirt at $24. What is the ratio of the money spent on a pair of socks to the total money spent? _____ _____

LESSON 120 Simplifying Ratios

A. One way to simplify ratios is to write them as fractions and simplify the fractions to the lowest terms. Then you can write the simplified fractions as ratios with a colon. Here are some examples of using this method. Complete the second example.

$12:20 = \dfrac{12}{20} = \dfrac{3}{5} = 3:5$ $35:42 = \dfrac{}{42} = \dfrac{5}{} = :$

B. Simplify each ratio.

18:27 =   24:28 =

16:32 =   42:54 =

27:15 =   20:56 =

60:90 =   44:72 =

72:84 =   33:88 =

42:210 =

24:108 =

Easy Peasy All-in-One Homeschool   EP Math 5/6 Workbook

LESSON 120 Practice

Simplify each ratio.

8:14 =   10:16 =

9:15 =   12:27 =

6:27 =   32:28 =

14:98 =   44:55 =

25:45 =   24:28 =

33:39 =   42:56 =

90:315 =

64:200 =

A math riddle for you! I am the ratio of the sum of two numbers whose product is 15 to their differences. What ratio am I?

LESSON 121 Simplifying Ratios

A. You can simplify ratios just as you simplify fractions: divide both numbers by their greatest common divisor (GCD) or keep dividing both sides by a common divisor until you can't divide them any further. The following examples use the strategy of 'keep dividing' to reach the simplest form of a ratio. Complete the second example.

÷3 (63 : 84) ÷3
   = 21 : 28
÷7 (   ) ÷7
   = 3 : 4

÷4 (128 : 200) ÷4
÷2 (   ) ÷2

B. Equivalent ratios are just like equivalent fractions. Ratios are equivalent if they can be simplified to the same ratio. To find equivalent ratios, multiply or divide both sides by the same number. Write at least three equivalent ratios for each ratio below.

                    x 2   ÷ 2
4:5   8:10, _____   20:50   10:25, _____

5:7   _____   60:90   _____

C. Simplify each ratio.

12:60   36:54   96:42

30:75   22:88   99:30

84:156   25:150   80:400

LESSON 121 Practice

A. Circle all ratios that are equivalent to the first one in each row.

12:15	4:5	36:45	3:5	45/36	5:4	24:30
36:27	12:6	4 to 3	16:12	3:4	72:54	6/8
40:32	20/18	10:8	4:5	15:12	44:36	5:4
44:66	22:33	3:2	11:22	2 to 3	4:6	55:77
25:50	1:2	12:25	5/10	4:10	10:20	2:5

B. Simplify each ratio.

45:50   54:72   58:20

42:63   75:21   16:64

54:108   48:132   44:121

105:375   320:608   125:350

LESSON 122 Word Problems: Equivalent Ratios

A. To solve word problems involving ratios, first identify the known ratio and the unknown ratio. Then use their relationship to find the unknown and finally what's being asked. Here is an example of using this strategy.

Jenny is having a party. She wants to prepare 1 pie for every 6 guests. If 24 guests are coming, how many pies does she need? _____

First, identify the known and unknown ratios:
   Known = Pie : Guests = 1:6
   Unknown = Pie : Guests = ? : 24
   Two ratios should be equivalent,
   so we need to find ? that makes 1:6 = ? : 24.

Then, find what is being asked:
   1 : 6 = 1 × 4 : 6 × 4 = ? : 24
   ? = 1 × 4 = 4

So, Jenny needs 4 pies.

B. Solve the word problems.

At a market, 5 tomatoes were being sold for $2. Peter bought $10 worth of tomatoes. How many tomatoes did he buy? _____

Sandy can solve 24 division problems every 20 minutes. How many problems can she solve in 10 minutes? _____

Ella planted 30 roses, 27 daisies, and 15 violets in her garden. For every 9 daisies, how many roses did Ella plant? _____

In a zoo, the ratio of adults to children is 5 to 7. If there are 125 adults in the zoo, how many children are there? _____

A recipe calls for 3 cups of milk for every 2 cups of flour. Sarah used 4 cups of flour. How many cups of milk did she use? _____

There are 12 girls in a book club, and the ratio of boys to girls is 2:1. How many members are in the club? _____

Rowan drew a rectangle whose length to width ratio was 3 to 2. If the width of the rectangle was 10 inches, what was its area? _____

A recipe calls for 2 cups of vegetables, 3 cups of chicken stock, and 2 teaspoons of salt to make 4 servings of soup. How many teaspoons of salt is used for every cup of vegetables? _____

Easy Peasy All-in-One Homeschool   EP Math 5/6 Workbook

LESSON 122 Practice

A. The table shows what Lisa planted in her garden. Use the table to answer the questions.

Plants in Lisa's Garden

Plant	# of plant
Peppers	20
Broccoli	18
Lettuce	30
Beets	12

For every 10 pepper plants, how many beet plants are there? _____

For every 6 broccoli plants, how many lettuce plants are there? _____

What is the ratio of pepper plants to all plants? _____

B. Solve the word problems.

Adam can solve 15 fraction problems every 18 minutes. How many problems can he solve in 6 minutes? _____

A restaurant makes cakes to pies in the ratio of 4 to 5. If the restaurant makes 35 pies, how many cakes does it make? _____

A recipe calls for 6 bananas to make 4 servings of banana pudding. How many bananas will be used to make 12 servings? _____

The ratio of boys to girls in a library is 5 to 6. If there are 20 boys in the library, how many girls are there? _____

In a grocery, apples and pears are sold in the ratio of 6:7. If the grocery sells 49 pears, how many apples are sold? _____

There are 12 boys in a math club, and the ratio of boys to girls is 3:4. How many members are in the club? _____

Mark drew a rectangle whose length to width ratio was 4 to 5. If the width of the rectangle was 15 inches, what was its perimeter? _____

Eight candies are sold at $2. Brian bought 40 candies and paid with $20. How much change did he receive? _____

Larry's car uses 5 gallons of gas to travel 110 miles. How many miles can Larry drive on one gallon of gas? _____

LESSON 123 Understanding Proportions

A. A **proportion** is an equation that states two ratios are equal. It can be written in two ways: in colon form, a:b = c:d, or in fraction form, a/b = c/d. Complete the examples.

$$3:4 = 15:20 \quad \text{or} \quad \frac{3}{4} = \frac{15}{20} \qquad 5:2 = 25:10 \quad \text{or} \quad \frac{5}{2} = \frac{25}{10}$$

B. A proportion is true if its ratios are equal. One way to determine if a proportion is true or false is to simplify the ratios and compare them, as shown in the example below.

$$\frac{6}{9} = \frac{12}{30} \quad \Rightarrow \quad \text{Reduce each side:} \quad \frac{2}{3} \neq \frac{2}{5} \quad \text{The ratios are not equal, so the proportion is false.}$$

C. Another way to determine if a proportion is true or false is to cross multiply the ratios and compare the cross products. Cross multiplying means to multiply diagonally across the equal sign, and cross products are the products found by cross multiplication. A proportion is true if the cross products are equal. Here is an example of using this method.

$$\frac{6}{15} \times \frac{10}{25} \quad \Rightarrow \quad \text{Cross multiply:} \quad \begin{array}{l} 6 \cdot 25 = 150 \\ 15 \cdot 10 = 150 \end{array} \quad \text{The cross products are equal, so the proportion is true.}$$

D. Determine whether each proportion is true or false. Show why.

6:7 = 12:14

32:40 = 10:8

5:2 = 20:15

8:14 = 16:28

$$\frac{9}{8} = \frac{15}{10}$$

$$\frac{24}{40} = \frac{16}{30}$$

$$\frac{18}{24} = \frac{6}{8}$$

$$\frac{14}{18} = \frac{35}{45}$$

Easy Peasy All-in-One Homeschool EP Math 5/6 Workbook

LESSON 123 Practice

A. Determine if each proportion is true or false. Show why.

5:9 = 15:45   15:10 = 12:8

7:4 = 63:36   21:49 = 7:14

$\dfrac{15}{45} = \dfrac{24}{64}$ $\dfrac{32}{40} = \dfrac{20}{25}$

$\dfrac{63}{36} = \dfrac{28}{16}$ $\dfrac{12}{44} = \dfrac{24}{66}$

$\dfrac{18}{54} = \dfrac{33}{99}$ $\dfrac{45}{40} = \dfrac{54}{48}$

$\dfrac{36}{48} = \dfrac{27}{36}$ $\dfrac{24}{30} = \dfrac{70}{54}$

B. Write a proportion for each situation. Write ratios using a colon.

Two bags of apples at $8; ? bags of apples at $40.   _____

4 free throws in 10 seconds; 20 free throws in ? seconds.   _____

255 miles on 10 gallons of gas; ? miles on 5 gallon of gas.   _____

80 miles per hour; 300 miles in ? hours.   _____

23 out of 25 problems; ? out of 100 problems.   _____

Easy Peasy All-in-One Homeschool   EP Math 5/6 Workbook

LESSON 124 Solving Proportions

A. What is cross multiplication? Explain how to determine if 4/10 = 6/15 is true using cross multiplication. Review Lesson 123 if needed.

B. Solving a proportion means finding an unknown part of the proportion. You can solve a proportion using cross multiplication. After cross multiplying, you divide each side by the number in front of the unknown part to find its value. Here is an example.

Cross multiply. Divide. Simplify.

$$\frac{?}{4} \times \frac{2}{5} \Rightarrow 5 \cdot ? = 4 \cdot 2 \Rightarrow \frac{\cancel{5} \cdot ?}{\cancel{5}} = \frac{4 \cdot 2}{5} \Rightarrow ? = \frac{8}{5} = 1\frac{3}{5}$$

C. Always reduce fractions before performing any operation. See how it gets simpler!

Reduce. Cross multiply. Divide. Simplify.

$$\frac{7}{?} \times \frac{\cancel{12}\,3}{\cancel{16}\,4} \Rightarrow 3 \cdot ? = 7 \cdot 4 \Rightarrow \frac{\cancel{3} \cdot ?}{\cancel{3}} = \frac{7 \cdot 4}{3} \Rightarrow ? = 9\frac{1}{3}$$

D. Solve each proportion. Reduce first if you can to make the numbers easier to work with. Simplify and convert your answers to mixed numbers.

$$\frac{?}{6} = \frac{5}{8} \qquad\qquad \frac{3}{?} = \frac{4}{3}$$

$$\frac{2}{9} = \frac{10}{?} \qquad\qquad \frac{4}{15} = \frac{?}{7}$$

$$\frac{?}{3} = \frac{10}{18} \qquad\qquad \frac{6}{?} = \frac{8}{12}$$

$$\frac{20}{25} = \frac{?}{20} \qquad\qquad \frac{?}{16} = \frac{24}{54}$$

Easy Peasy All-in-One Homeschool EP Math 5/6 Workbook

LESSON 124 Practice

A. Solve each proportion using cross multiplication. Simplify and convert your answers to mixed numbers.

$$\frac{?}{4} = \frac{7}{8}$$

$$\frac{6}{?} = \frac{2}{7}$$

$$\frac{3}{8} = \frac{4}{?}$$

$$\frac{2}{9} = \frac{?}{3}$$

$$\frac{6}{?} = \frac{9}{11}$$

$$\frac{8}{?} = \frac{7}{12}$$

$$\frac{5}{4} = \frac{25}{?}$$

$$\frac{13}{20} = \frac{?}{6}$$

B. Reduce first, then solve using cross multiplication. Write your answers in simplest form.

$$\frac{?}{8} = \frac{14}{21}$$

$$\frac{16}{10} = \frac{6}{?}$$

$$\frac{15}{?} = \frac{18}{45}$$

$$\frac{30}{?} = \frac{66}{77}$$

$$\frac{36}{60} = \frac{21}{?}$$

$$\frac{16}{40} = \frac{?}{16}$$

LESSON 125 Solving Proportions

A. Explain how to solve a proportion using cross multiplication. Use the proportion below to show your steps. Review Lesson 124 if needed.

$$\frac{11}{?} \times \frac{5}{12} \Rightarrow \text{Cross multiply.} \Rightarrow \text{Divide.} \Rightarrow \text{Simplify.}$$

B. Solve each proportion. Reduce first if you can to make the numbers easier to work with. Simplify and convert your answers to mixed numbers.

$$\frac{?}{4} = \frac{4}{7} \qquad\qquad \frac{8}{?} = \frac{6}{9}$$

$$\frac{8}{12} = \frac{10}{?} \qquad\qquad \frac{?}{14} = \frac{5}{42}$$

$$\frac{15}{28} = \frac{10}{?} \qquad\qquad \frac{?}{25} = \frac{12}{45}$$

$$\frac{18}{63} = \frac{?}{14} \qquad\qquad \frac{40}{25} = \frac{32}{?}$$

C. Write a proportion and solve it to answer each question.

Six apples are sold at $5.25. How much do 18 apples cost?

Angie drove 55 miles per hour for 4 hours. How many miles did she drive?

Six cans of soda cost $2.50. How many cans of soda can you buy for $25?

Josh drove 156 miles at 52 miles per hour. How many hours did he drive?

Easy Peasy All-in-One Homeschool

EP Math 5/6 Workbook

LESSON 125 Practice

Solve each proportion. Reduce first if you can to make the numbers easier to work with. Simplify and convert your answers to mixed numbers.

$\dfrac{?}{6} = \dfrac{4}{5}$ $\qquad\qquad\qquad$ $\dfrac{3}{4} = \dfrac{?}{9}$

$\dfrac{?}{12} = \dfrac{3}{8}$ $\qquad\qquad\qquad$ $\dfrac{7}{?} = \dfrac{4}{10}$

$\dfrac{8}{36} = \dfrac{?}{9}$ $\qquad\qquad\qquad$ $\dfrac{8}{20} = \dfrac{?}{6}$

$\dfrac{?}{4} = \dfrac{13}{22}$ $\qquad\qquad\qquad$ $\dfrac{12}{18} = \dfrac{14}{?}$

$\dfrac{40}{25} = \dfrac{9}{?}$ $\qquad\qquad\qquad$ $\dfrac{?}{16} = \dfrac{15}{28}$

$\dfrac{14}{28} = \dfrac{21}{?}$ $\qquad\qquad\qquad$ $\dfrac{11}{24} = \dfrac{?}{20}$

$\dfrac{?}{30} = \dfrac{20}{16}$ $\qquad\qquad\qquad$ $\dfrac{15}{21} = \dfrac{?}{49}$

$\dfrac{17}{?} = \dfrac{30}{25}$ $\qquad\qquad\qquad$ $\dfrac{?}{16} = \dfrac{18}{30}$

Easy Peasy All-in-One Homeschool $\qquad\qquad\qquad$ EP Math 5/6 Workbook

LESSON 126 Finding Unit Rates

A. A **rate** is a ratio of two quantities with different units. A **unit rate** is a rate with a denominator of 1. Unit rates are often written with a slash (/) or the word "per." Here is an example of a rate and its unit rate written in three different ways.

$$\frac{120 \text{ miles}}{3 \text{ hours}} = \frac{120 \div 3}{3 \div 3} = \frac{40 \text{ miles}}{1 \text{ hour}} \quad \text{or} \quad \begin{array}{l} 40 \text{ miles/hour} \\ 40 \text{ miles } \textbf{per} \text{ hour} \end{array}$$

B. To find a unit rate, first write the rate as a fraction. Then divide both sides of the rate by the denominator. Here is an example of converting a rate to a unit rate.

I can walk 18 miles in 4 hours. $\Rightarrow \dfrac{18 \text{ miles}}{4 \text{ hours}} = \dfrac{18 \div 4}{4 \div 4} = \begin{array}{l} 4.5 \text{ miles/hour} \\ 4.5 \text{ miles } \textbf{per} \text{ hour} \end{array}$

C. Find a unit rate to represent each situation.

Buy 4 tickets at $126.
How much per ticket?

Run 35 miles every 7 days.
How many miles per day?

Type 675 words in 15 minutes.
How many words per minute?

Need 3 eggs to bake 24 cookies
How many cookies per egg?

D. Solve the word problems. You may use a calculator.

Mia can swim 20 meters per minute. If she swims 30 minutes each day, how many days will it take her to swim 9,000 meters? _____

Sam is painting 1,200 square feet of wall space. Paint costs $26 per gallon, and one gallon covers 300 square feet. How much would it cost Sam to paint the entire wall space? _____

Joe had a road trip across the United States during the summer. He drove 5,850 miles for 18 days at an average speed of 65 miles per hour. How many hours did he drive per day? _____

Easy Peasy All-in-One Homeschool   EP Math 5/6 Workbook

LESSON 126 Practice

A. Find a unit rate to represent each situation.

2 inches of rain in 4 hours.
How many inches per hour?

216 miles on 9 gallons of gas.
How many miles per gallon?

20 video lessons over 5 days.
How many lessons per day?

195 heartbeats in 3 minutes.
How many heartbeats per minute?

5 laps in 15 minutes.
How many minutes per lap?

150 customers during 6 hours.
How many customers per hour?

B. Solve the word problems. You may use a calculator.

Ana can type 45 words per minute. If she types 20 minutes each day, how many days will it take her to type 4,500 words? _____

A bakery bakes 3 trays of cookies every 2 hours. Each tray holds 36 cookies. How many cookies does the bakery bake in 6 hours? _____

A company has 4 factories. Each factory produces 8 cars per hour. All factories operate 8 hours a day and 5 days a week. How many cars does the company produce in 4 weeks? _____

Kim has completed 16 worksheets of 25 math problems each in 20 days. On average it took her 3 minutes per problem. How many hours per day did she spend on math problems? _____

Max is driving for 4.5 hours today. He wants to drive at a speed of 56 miles per hour. His car can go 28 miles per gallon of gas. How many gallons of gas does he need for his journey? _____

Easy Peasy All-in-One Homeschool   EP Math 5/6 Workbook

LESSON 127 Finding Unit Prices

A. A **unit price** is a price per unit. It is often used to compare the prices of different sizes (or brands) of the same product. One way to find a unit price is to use a proportion.

A 3-pound bag of sugar for $3.60 ⇨ $\dfrac{\$3.60}{3\ pounds} = \dfrac{?}{1\ pound}$ ⇨ $? = \dfrac{\$3.60}{3} = \$1.20$

B. A shortcut method to find a unit price is simply to divide the price by the quantity. Here is an example of comparing unit prices to determine the better buy.

| A 2-pound bag of flour for $1.40 | or | A 5-pound bag of flour for $2.50 |

$\dfrac{\$1.40}{2\ pounds} = \$0.70/pound$ \qquad $\dfrac{\$2.50}{5\ pounds} = \$0.50/pound$ (circled)

C. Find the unit price of each item. Determine the better buy for each pair.

2 pounds of beef for $10.90 \qquad or \qquad 4 pounds of beef for $18.80

12 inches of ribbon for $3 \qquad or \qquad 36 inches of ribbon for $7.20

1 gallon of paint for $26 \qquad or \qquad 5 gallons of paint for $150

A dozen eggs for $1.80 \qquad or \qquad Two dozen eggs for $4.80

6 bottles of soda for $5.40 \qquad or \qquad 12 bottles of soda for $12.00

Easy Peasy All-in-One Homeschool \hfill EP Math 5/6 Workbook

LESSON 127 Practice

A. Find the unit price of each item. Circle the better buy between each pair.

    2 kilograms of rice for $6.90   or   5 kilograms of rice for $18.00

    8 ounces of honey for $5.20   or   16 ounces of honey for $9.60

    10 meters of wire for $15   or   50 meters of wire for $80

    2 cucumbers for $1.10   or   15 cucumbers for $7.50

B. Go to a grocery store. Write the price and unit price as well as the measurements (for example, 4 ounces) for five different products. At home see if you can take the product price and measurements and come up with the same unit price as the store.

The information on the tag				Your calculation
Product name	Price	Measurements	Unit price	Unit price

Easy Peasy All-in-One Homeschool   EP Math 5/6 Workbook

LESSON 128 Converting Units

A. Proportions can be used to convert between different units. Here is an example of using a proportion to convert feet to inches. Complete the example.

There are 12 inches in 1 foot. How many inches are in 8 feet?

$$\frac{12 \text{ inches}}{1 \text{ foot}} \bowtie \frac{? \text{ inches}}{8 \text{ feet}} \quad \Rightarrow \quad 1 \cdot ? = 12 \cdot 8 \quad \Rightarrow \quad ? =$$

B. Another way to convert units is to multiply ratios called **conversion factors**. Conversion factors are relationships between two units. The goal is to cancel out the units you don't want and leave the units you do. Here are some examples of using conversion factors to convert units. Complete the examples.

There are 1000 meters in 1 kilometer. What is 5800 meters in kilometers?

$$5800 \; \cancel{\text{meters}} \times \frac{1 \text{ kilometer}}{1000 \; \cancel{\text{meters}}} = \underline{} \text{ kilometers}$$

What is 10 m/s (meters per second) in km/h (kilometers per hour)?

$$\frac{10 \; \cancel{\text{meters}}}{1 \; \cancel{\text{second}}} \times \frac{1 \text{ kilometer}}{1000 \; \cancel{\text{meters}}} \times \frac{3600 \; \cancel{\text{seconds}}}{1 \text{ hour}} = \underline{} \text{ km/h}$$

C. Convert each unit as indicated.

There are 5,280 feet in a mile.
What is 2.5 miles in feet?

What is 60 mph (miles per hour) in fps (feet per second)?

There are 1,609 meters in a mile.
What is 3218 meters in miles?

What is 440 fps (feet per second) in mph (miles per hour)?

One inch is 2.54 centimeters.
What is 5 inches in centimeters?

What is 180 mph (miles per hour) in m/s (meters per second)?

Lesson 128 Practice

Convert each unit as indicated.

What is 1,200 meters in kilometers?   What is 1.2 miles in feet?

What is 804.5 meters in miles?   What is 8 inches in centimeters?

What is 1,320 feet in miles?   What is 0.04 kilometer in meters?

What is 22.86 centimeters in inches?   What is 0.3 mile in meters?

There are 3 feet in 1 yard.   There are 100 centimeters in 1 meter.
What is 15 feet in yards?   What is 75,000 centimeters in kilometers?

What is 50 m/s (meters per second)   What is 72 mph (miles per hour)
in km/h (kilometers per hour)?   in m/s (meters per second)?

What is 360 km/h (kilometers per hour)   What is 4,400 fps (feet per second)
in m/s (meters per second)?   in mph (miles per hour)?

Easy Peasy All-in-One Homeschool   EP Math 5/6 Workbook

LESSON 129 Converting Units of Length

A. Here are the conversion charts for units of length.

Customary units of length:

1 mile (mi) = 1,760 yards (yd)
1 mile (mi) = 5,280 feet (ft)
1 yard (yd) = 3 feet (ft)
1 foot (ft) = 12 inches (in)

Metric units of length:

1 kilometer (km) = 1000 meters (m)
1 meter (m) = 100 centimeters (cm)
1 centimeter (cm) = 10 millimeters (mm)

Customary to metric:

1 mi = 1.61 km
1 yd = 0.91 m
1 ft = 30.48 cm
1 in = 2.54 cm

Metric to customary:

1 km = 0.625 mi
1 m = 1.1 yd
1 cm = 0.033 ft
1 cm = 0.396 in

B. Let's look at an example. How many centimeters are in 5 yards? You can first convert yards to meters and then convert meters to centimeters.

$$\frac{1\ yd}{0.91\ m} = \frac{5\ yd}{?\ m} \quad \begin{array}{l} ? = 5 \cdot 0.91 \\ ? = 4.55\ m \end{array} \Rightarrow \frac{1\ m}{100\ cm} = \frac{4.55\ m}{?\ cm} \quad ? = 455\ cm$$

C. Convert each unit as indicated. Round off decimals to 2 decimal places, if necessary. (Note that you are using approximate values. Depending on how you convert, your answers may be off a bit from the answer key. If your answers are close, you are correct!)

12 yd ⇨ ft

5.8 km ⇨ m

98 in ⇨ ft and in

0.037 m ⇨ mm

52 ft ⇨ yd and ft

468,000 cm ⇨ km

2 yd ⇨ cm

10 m ⇨ ft

Easy Peasy All-in-One Homeschool

EP Math 5/6 Workbook

LESSON 129 Practice

Convert each unit as indicated. Round off decimals to 2 decimal places, if necessary. (Note that you are using approximate values. Depending on how you convert, your answers may be off a bit from the answer key. If your answers are close, you are correct!)

4 ft 6 in ⇨ in   0.7 m ⇨ mm

2.5 mi ⇨ yd   2 m 4 cm ⇨ cm

58 in ⇨ ft and in   1290 m ⇨ km

76 ft ⇨ yd and ft   8,300 mm ⇨ m

6 mi ⇨ m   5 km ⇨ mi

0.5 in ⇨ mm   60 cm ⇨ in

500 yd ⇨ km   30 m ⇨ ft

20,000 in ⇨ km   7 km ⇨ mi

LESSON 130 Customary Units and Metric Units

A. There are two major systems of measurement in use today: the **customary system** and the **metric system**. The customary system is used in the United States while the metric system is used in most other countries. The metric units are used in math and science because they are based on multiples of 10, which makes them easy to convert.

B. Sort and record the following units of measurement into the table.

Pound	Liter	Meter	Ounce	Foot	Cup
Kilogram	Gallon	Quart	Centimeter	Pint	Millimeter
Yard	Kilometer	Mile	Gram	Milliliter	Inch

	Length	Weight/mass	Volume/capacity
Customary units			
Metric units			

C. When in use, the units are usually abbreviated as symbols. Find all pairs that incorrectly match the symbol and unit of measure. (Hint: There are 7 wrong pairs.)

cm – centimeter	mi – meter	gal – gram	kg – kilogram
lb – pound	L – liter	oz – ounce	qt – pint
g – gallon	in – inch	mL – mile	yd – yard
km – kilometer	pt – quart	ft – foot	m – millimeter

Easy Peasy All-in-One Homeschool — EP Math 5/6 Workbook

LESSON 130 Practice

A. Get your ruler. Measure an object in centimeters and inches. Record them as a ratio.

Object	In inches	In centimeters	Ratio

B. Measure two other objects just in inches. Set up proportions using the above ratio. Convert inches to centimeters. Now measure them in centimeters to see if you are right.

Object	Measure in inches and convert to centimeters	Actual length in centimeters

C. Write the best unit of length for each measurement. Answers may be repeated.

I want to measure …	Customary unit	Metric unit
The distance between two cities		
The width of my math workbook		
The length of my room		
The length of a piece of fabric		

D. Draw lines to match the measurement tools with the units they measure in.

Scale   Thermometer   Ruler   Protractor   Beaker

Inches   Grams   Milliliters   Celsius   Degrees

Easy Peasy All-in-One Homeschool   EP Math 5/6 Workbook

LESSON 131 Converting between Metric Units

A. The metric system is a decimal-based system, meaning the units are defined in multiples of 10 from the base unit. The prefixes are used to indicate which multiple of 10 is being used. As a result, converting between metric units is the same as switching between prefixes, which is simply moving the decimal point. The table below shows metric prefixes, symbols, and their meanings. The most common base units are the meter, gram, and liter.

Divide by 10s or move the decimal point to the left.

Prefix	Kilo-	Hecto-	Deka-	Base unit	Deci-	Centi-	Milli-
Symbol	k	h	da	-	d	c	m
Meaning	1000	100	10	1	0.1	0.01	0.001

Multiply by 10s or move the decimal point to the right.

B. Make your own chart for commonly used metric units. Convert each unit to the other units to show how many are in each other. The first one is done for you!

Units of length: kilometers (km), meters (m), centimeters (cm), and millimeters (mm)

1 km = 1,000 m = 100,000 cm = 1,000,000 mm

1 m =

1 cm =

1 mm =

Units of weight: kilograms (kg), grams (g), milligrams (mg)

1 kg =

1 g =

1 mg =

Units of volume: kiloliters (kL), liters (L), milliliters (mL)

1 kL =

1 L =

1 mL =

LESSON 131 Practice

A. Convert each unit as indicated.

0.5 kg =	g	0.17 m =	mm
45 cm =	mm	0.08 km =	cm
0.89 L =	mL	3,360 mg =	g
2300 g =	kg	0.00027 kL =	mL
620 cm =	m	500,000 mm =	km

B. Which is longer, heavier, or larger in each pair?

810 mm or 8.1 cm	120 kL or 1,200 L
240 mL or 2.4 L	0.09 m or 900 mm
300 mg or 30 g	180 cm or 0.018 m
0.05 m or 5 mm	0.17 kg or 17,000 g

C. Solve the word problems.

Emma runs 300 meters per day to stay fit. How many kilometers does she run in 4 weeks? _____

Sunny had a ribbon 2.5 m long. She cut some off and 85 cm was left. How much did Sunny cut off? _____

One bucket holds 2.5 liters of water. Walter needs 5 buckets of water for his garden. How much water does he need in liters? _____

Joan had a cold. She took 4 ml of medicine, 3 times a day, for 5 days. How much medicine did she take? _____

Peter bought a 2 kg bag of flour. He used 460 grams of it to bake cookies and 780 grams to bake pies. How much flour does he have left? _____

LESSON 132 Converting Customary Units of Volume

A. Here is the conversion chart for customary units of volume.

Customary units of volume:

1 gallon (gal) = 4 quarts (qt)
1 quart (qt) = 2 pints (pt)
1 pint (pt) = 2 cups (c)
1 cup (c) = 8 fluid ounces (fl oz)
1 cup (c) = 16 tablespoons (tbsp)
1 tablespoon (tbsp) = 3 teaspoons (tsp)

A gallon contains:

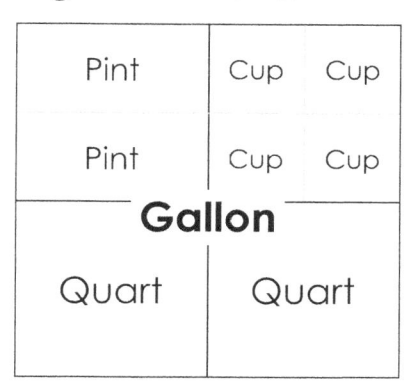

Pint	Cup	Cup
Pint	Cup	Cup
Gallon		
Quart	Quart	

B. Convert each unit as indicated.

8 gal ⇨ pt   18 qt ⇨ gal

9 qt ⇨ c   64 fl oz ⇨ qt

44 c ⇨ pt   24 pt ⇨ gal

C. Solve the word problems.

A recipe calls 2 cups of milk to make a dozen cookies. How many cookies can you make using a gallon of milk? _____

Dana drinks a cup of milk every day. How many fluid ounces of milk does she drink in one week? _____

A bucket had 2.5 gallons of water. Kyle used some, and 7 quarts of water are left. How many quarts of water did Kyle use? _____

Matt bought a pack of 24 soda cans. Each can contains 12 fl oz of soda. How many pints of soda are in the pack? _____

Easy Peasy All-in-One Homeschool   EP Math 5/6 Workbook

LESSON 132 Practice

A. Convert each unit as indicated.

150 qt ⇨ gal   256 c ⇨ gal

64 pt ⇨ qt   5.25 gal ⇨ qt

3 gal ⇨ c   48 pt ⇨ gal

1024 c ⇨ qt   350 qt ⇨ pt

60 pt ⇨ c   120 c ⇨ qt

B. Solve the word problems.

A recipe calls for 2 quarts of milk, but Josh has only 2 cups of milk. How much more milk does he need in cups? _____

Olivia filled up 20 fl oz water bottles with 5 pints of water. How many bottles did she fill up? _____

There are 6 quarts of water in a jug, and 8 more pints of water will fill it up completely. What is the capacity of the jug in gallons? _____

Container #1 holds 1,800 pints of water. Container #2 holds 200 gallons of water. Which one holds more water? _____

LESSON 133 Converting Customary Units of Weight

A. Here are the conversion charts for units of weight.

Customary units of weight:

1 ton (t) = 2,000 pounds (lb)
1 pound (lb) = 16 ounces (oz)

Customary to metric:

1 t = 907.2 kg
1 lb = 453.59 g

B. Convert each unit as indicated. The first one is done for you!

6 lb ⇨ oz

$\dfrac{1\ lb}{16\ oz} = \dfrac{6\ lb}{?\ oz}$? = 6 × 16 = 96 oz

0.02 t ⇨ oz

64 oz ⇨ lb

8,000 lb ⇨ t

0.7 t ⇨ lb

32,000 oz ⇨ t

5 t ⇨ kg

6 lb ⇨ g

C. Solve the word problems.

Stacey bought four dozen cookies. Each cookie weighs 1.5 ounces. How many pounds of cookies did she buy? _____

Daniel bought a cake that weighed 4 pounds and cut it into 8 equal pieces. How many ounces does each piece weigh? _____

A book store found that the average weight of their books was 12 ounces. What will be the weight of 2,000 books in pounds? _____

A farmer packaged 2 tons of his beans into 200 2-pound bags and 300 5-pound bags and sold them to a local grocery. How many pounds of the beans does he have left after that? _____

Easy Peasy All-in-One Homeschool   EP Math 5/6 Workbook

LESSON 133 Practice

A. Convert each unit as indicated. Round off decimals to 2 decimal places, if necessary.

4.1 lb ⇨ oz0.1 t ⇨ oz

72,000 oz ⇨ t2 lb ⇨ g

8 lb ⇨ kg152 oz ⇨ lb

32 oz ⇨ g0.062 t ⇨ lb

2 t ⇨ kg400 lb ⇨ t

B. Solve the word problems.

A pack of 12 pencils weighs 1 pound 2 ounces. How many ounces does each pencil weigh? _____

Sam's cat weighs 6 pounds 8 ounces. Laura's cat weighs 12 ounces less than Sam's cat. How much does Laura's cat weigh? _____

Jim harvested 9 tons of corn and loaded it into trucks that can carry 6,000 pounds each. How many trucks did he load? _____

Mark put 128 candies equally into 8 boxes. Each candy weighs 4 ounces. How many pounds does each box weigh? _____

Easy Peasy All-in-One HomeschoolEP Math 5/6 Workbook

LESSON 134 Converting Temperatures

A. The most commonly used temperature scales are the **Fahrenheit** scale (°F) and the **Celsius** scale (°C). The United States mainly uses the Fahrenheit scale while most of the world uses the Celsius scale.

B. Here are the formulas to convert between Fahrenheit and Celsius temperatures.

Celsius to Fahrenheit

$$°F = \frac{9}{5}°C + 32$$

Fahrenheit to Celsius

$$°C = \frac{5}{9}(°F - 32)$$

C. Convert Fahrenheit to Celsius and Celsius to Fahrenheit. Round off decimals to 1 decimal place, if necessary.

Human body: 98.6 °F

Summer at South Pole: -18 °F

Room temperature: 20 °C

Winter at South Pole: -60 °C

Water's freezing point: 0 °C

Summer day on Mars: 70 °F

Water's boiling point: 212 °F

Summer night on Mars: -73 °C

Easy Peasy All-in-One Homeschool

EP Math 5/6 Workbook

LESSON 134 Practice

Convert each temperature. Round off decimals to 1 decimal place, if necessary.

Absolute zero: -273 °C

Steel's melting point: 2500 °F

Dog's body: 102 °F

Summer at North Pole: 32 °F

Chicken's body: 41.7 °C

Winter at North Pole: -40 °C

Earth's core: 10,800 °F

Hottest weather recorded: 134 °F

Surface of sun: 5,505 °C

Coldest weather recorded: -89.2 °C

Did you know? There is another temperature scale used in scientific measurements: the **Kelvin** scale (K). It is named in honor of the Scottish physicist who first defined it. Its zero point, 0 K, is defined to coincide with the coldest physically possible temperature, called absolute zero, at which nearly all molecules stop moving. The temperature of absolute zero is 0 K = -459.67 °F = -273.15 °C.

LESSON 135 Converting Customary and Metric Units

Answer each question in the unit used in your country and then convert. Use the conversion charts in Lessons 129 through 134.

How tall are you in inches and in centimeters?

How long is your foot in inches and in millimeters?

How long is your living room in feet and in meters?

Estimate how much water you drink per day in cups and in liters? (1 cup = 0.24 L)

How much do you weigh in pounds and in kilograms? (1 kg = 2.2 lb)

How much did you weigh when you were born in pounds and in kilograms?

What is the temperature today in Fahrenheit and in Celsius?

How old are you in years, in months, and in days?

Lesson 135 Practice

Determine the better buy for each pair.

2 feet of ribbon for $3   or   2 meters of ribbon for $5

2 gallons of milk for $4   or   200 milliliters of milk for $1

4 pounds of beef for $16   or   300 grams of beef for $4

A 2-liter bottle of soda for $2   or   Six 12-ounce cans of soda for $2

Five 2-pound bags of sugar for $11   or   Two 1-kilogram bags of sugar for $6

LESSON 136 Exponents with Integer Bases

A. Earlier in the course you learned to multiply negative numbers: multiplying a positive by a negative is a negative, and multiplying two negatives is a positive. What if you multiply more than two negatives? If there is an even number of negatives, the answer will be positive. If there is an odd number of negatives, the answer will be negative. Check the signs in the examples below. Review Lesson 31 if needed.

-2 × -2 = 4   2 negatives make a positive.

-2 × -2 × -2 = 4 × -2 = -8   3 negatives make a negative.

-2 × -2 × -2 × -2 = 4 × 4 = 16   4 negatives make a positive.

B. You also learned that the expression b^n means the number b is multiplied n times. Note that the number b can be any number including negative numbers. Study the examples carefully. Pay attention to the signs of the numbers. Review Lessons 31 and 32 if needed.

$3^2 = 3 \cdot 3 = 9$ $\qquad\qquad\qquad$ $4^3 = 4 \cdot 4 \cdot 4 = 64$

$(-3)^2 = -3 \times -3 = 9$ $\qquad\qquad$ $(-4)^2 = -4 \times -4 = 16$

$(-3)^3 = -3 \times -3 \times -3$ $\qquad\qquad$ $(-4)^4 = -4 \times -4 \times -4 \times -4$

$\qquad\quad = 9 \times -3 = -27$ $\qquad\qquad\quad\;\; = 16 \times 16 = 256$

C. Evaluate each expression. Write your answers without exponents. Simplify fractions.

$1^6 =$ $\qquad\qquad\qquad\qquad\qquad$ $0^3 =$

$2^5 =$ $\qquad\qquad\qquad\qquad\qquad$ $(-4)^2 =$

$12^2 =$ $\qquad\qquad\qquad\qquad\qquad$ $(-1)^9 =$

$10^2 =$ $\qquad\qquad\qquad\qquad\qquad$ $(-100)^3 =$

$7^1 \cdot 3^2 =$ $\qquad\qquad\qquad\qquad\;\,$ $(-1)^7 \cdot 2^6 =$

$5^2 / 10^2 =$ $\qquad\qquad\qquad\qquad\;\,$ $9^1 / (-6)^2 =$

LESSON 136 Practice

A. Multiply the numbers.

-9 × 8 =   7 × -7 =   -7 × 1 =

7 × -5 =   -6 × -4 =   -8 × -7 =

-5 × -9 =   7 × 2 =   3 × -6 =

3 × 4 =   -5 × 8 =   -9 × 0 =

B. Rewrite each expression using exponents.

5 × 5 × 5 =   -1 × -1 × -1 × -1 × -1 =

125 × 125 × 125 × 125 × 125 × 125 × 125 × 125 =

-43 × -43 × -43 × -43 × -43 × -43 × -43 × -43 × -43 =

C. Evaluate each expression. Write your answers without exponents. Simplify fractions.

$0^7 =$ $\qquad$ $(-1)^3 =$

$1^5 =$ $\qquad$ $(-5)^2 =$

$2^5 =$ $\qquad$ $(-2)^7 =$

$3^4 =$ $\qquad$ $(-4)^3 =$

$20^3 =$ $\qquad$ $(-100)^1 =$

$4^2 \cdot 5^2 =$ $\qquad$ $(-1)^5 \cdot 2^3 =$

$6^1 \cdot 4^1 =$ $\qquad$ $7^2 \cdot (-1)^8 =$

$1^5 / 7^2 =$ $\qquad$ $3^4 / (-3)^2 =$

$6^1 / 3^3 =$ $\qquad$ $(-2)^3 / (-4)^2 =$

Easy Peasy All-in-One Homeschool   EP Math 5/6 Workbook

LESSON 137 Patterns in Zeros

A. A **power of 10** is 10 multiplied by itself a certain number of times. To calculate a power of 10, simply put as many zeros as the exponent after the 1. Review Lesson 100 if needed.

$10^3 = 10 \cdot 10 \cdot 10 = 1{,}000$ $10^8 = 100{,}000{,}000$ (8 zeros)

B. To multiply by a power of 10, simply move the decimal point to the right as many places as the exponent. Remember that whole numbers have an implied decimal point at the end. See the examples below. Review Lesson 100 if needed.

$1.7 \times 10^4 = 1.7 \times 10000 = 17{,}000$ Move 4 places to the right.

$38 \times 10^5 = 38 \times 100000 = 3{,}800{,}000$ Move 5 places to the right.

C. Division works the same way except you move the decimal point in the opposite direction to the left. Review Lesson 102 if needed.

$1.7 \div 10^4 = 1.7 \div 10000 = 0.00017$ Move 4 places to the left.

$38 \div 10^5 = 38 \div 100000 = 0.00038$ Move 5 places to the left.

D. Multiply or divide by a power of 10.

$0.302 \times 10^2 =$ $0.3 \div 10^2 =$

$34.2 \div 10^3 =$ $2.51 \times 10^4 =$

$1{,}746 \times 10^3 =$ $520 \times 10^5 =$

$712 \div 10^5 =$ $5{,}216 \div 10^3 =$

E. Fill in the blank to make each statement true.

$7.34 \div \underline{} = 0.00734$ $\underline{} \times 10^2 = 15$

$246 \times \underline{} = 24{,}600$ $\underline{} \div 10^3 = 9{,}200$

$0.38 \times \underline{} = 3{,}800$ $\underline{} \times 10^5 = 673$

Easy Peasy All-in-One Homeschool   EP Math 5/6 Workbook

LESSON 137 Practice

A. Evaluate each expression. How many zeros are in each number?

$10^4 =$ $10^7 =$

B. Multiply or divide by a power of 10.

$64.11 \div 10^3 =$ $0.03 \div 10^2 =$

$16.3 \times 10^2 =$ $0.519 \times 10^4 =$

$3,106 \times 10^4 =$ $4,058 \div 10^5 =$

$23,774 \div 10^4 =$ $51,704 \times 10^3 =$

C. Fill in the blank to make each statement true.

$0.01 \times \underline{} = 1,000$ $\underline{} \times 10^4 = 720$

$3,109 \times \underline{} = 310,900$ $\underline{} \div 10^7 = 0.008$

$64,500 \div \underline{} = 0.0645$ $\underline{} \times 10^2 = 41,500$

D. What number am I? Use the clues to solve each riddle.

I am a percent. In fraction form, I am equivalent to three quarters minus two eighths. What number am I?   _____

I am a percent less than 100%. As a decimal, I have two digits in total and my tenths digit is the square of 3. What number am I? _____

I am a fraction. My numerator is 8 less than my denominator. In percent form, I can be written as 60%. What number am I?   _____

I am a decimal with 2 digits in total. As a percent, I'm between 200% and 300%. The sum of my digits is 5. What number am I?   _____

A math riddle for you! Which is heavier? A pound of iron or a pound of feathers?

Easy Peasy All-in-One Homeschool   EP Math 5/6 Workbook

LESSON 138 Order of Operations with Exponents

A. Evaluate each expression. Show your steps clearly in your notebook. These are the examples shown in Lessons 42 and 52. Review those lessons if needed.

$7 - 3 \times 5 + 6 =$

$3 + 2 \times \text{-}8 \div 4 - 5 =$

$5 + 2 \times (3 + 5) =$

$9 \div 3 \times (5 - 7) =$

B. When an expression has exponents, evaluate them after parentheses. Remember the order PEMDAS! Complete the third and fourth examples.

$9 \div 3^2 \times (5 - 7)$	Parentheses
$= 9 \div 3^2 \times \text{-}2$	Exponent
$= 9 \div 9 \times \text{-}2$	Division, then Multiplciation (left to right)
$= 1 \times \text{-}2$	
$= \text{-}2$	

$6 \times (5 \times 2 - 2^3)$	Parentheses (E)
$= 6 \times (5 \times 2 - 8)$	Parentheses (M)
$= 6 \times (10 - 8)$	Parentheses (S)
$= 6 \times 2$	Multiplication
$= 12$	

$4^2 + (3 \times 2^2)^2$	Parentheses (E)
$=$	Parentheses (M)
$=$	All exponents
$=$	Addition
$=$	

$(\text{-}9)^2 \div (5 - 8)^3$	Parentheses (S)
$=$	All exponents
$=$	Division
$=$	

D. Evaluate each expression using the order of operations. Do the left column first.

$2^3 + 4 \times 3^2 =$

$1^3 \times (\text{-}5 - 1)^2 \div 2^2 =$

$(\text{-}6)^2 \div 9 \times 2 =$

$(5 + 2^2 \div 1) - (\text{-}1)^5 =$

$(2^3 + 2) \times \text{-}5 =$

$2 \times (3 + 3^2) \div 6 - 3^2 =$

$(5 + 9)^2 \div 7^2 =$

$(8 \times 5 - 7 \times 6)^5 \div 4^2 =$

Easy Peasy All-in-One Homeschool   EP Math 5/6 Workbook

LESSON 138 Practice

Evaluate each expression using the order of operations.

$3 - 2 - 7 =$

$8 \times 6 \div 12 \times 4 \div -8 =$

$5 + 8 \div 4 =$

$3 \times 5 + -16 \div 2 - 9 =$

$-4 \times 6 - 12 =$

$9 + 20 \div -5 \times 4 - 3 =$

$8 - 35 \div -5 =$

$4 \times 9 - 72 \div 9 \div -4 =$

❄ ❄ ❄ ❄ ❄ ❄ ❄ ❄ ❄

$5 + (4 - 6) =$

$5 \times (3 + 2) \times 4 =$

$9 \div (6 \div 2) =$

$(7 + 9) \div 2 \times -6 =$

$(-5 + 8) \times 5 =$

$7 + 3 \times (-6 - 6) \div -3 =$

$-4 \times (3 - 9) =$

$(-5 + 45 \div 9 \times -4) - 5 =$

❄ ❄ ❄ ❄ ❄ ❄ ❄ ❄ ❄

$2^3 - 3^2 + 5 =$

$(3 - 6) \times (5 - 5^2) =$

$6^2 + 7 \times 2^3 =$

$(6 - 2^3 \times 2)^3 \div 10^2 =$

$5^2 \times 4 \div 2^2 =$

$-2 \times (4 - 8)^2 \div 4 - 1^4 =$

Easy Peasy All-in-One Homeschool

EP Math 5/6 Workbook

LESSON 139 Understanding Square Roots

A. A **square** of a number is the number times itself: $x^2 = x \cdot x$. The name "square" comes from the fact that the area of a square of side length x is equal to x^2, as shown below.

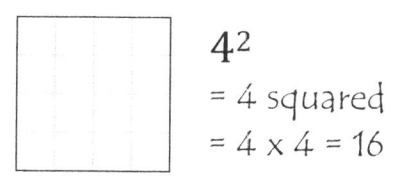

2^2
= 2 squared
= 2 × 2 = 4

3^2
= 3 squared
= 3 × 3 = 9

4^2
= 4 squared
= 4 × 4 = 16

B. The **square root** of a number n is a number x whose square is the given number n: $x^2 = n$. It is just the opposite of the square, represented by the symbol $\sqrt{}$: $\sqrt{n} = x$. The following examples show the relationship between square and square root. Complete the examples.

3 squared is 9.

$3^2 = 9$
$\sqrt{9} = 3$

The square root of 9 is 3.

15 squared is 225.

$15^2 = 225$
$\sqrt{225} =$

The square root of 225 is?

64 squared is?

$64^2 =$
$\sqrt{4096} = 64$

The square root of 4096 is 64.

C. Find the square root of each number. (Hint: Ask yourself, "What number multiplied by itself will produce this number?")

$\sqrt{1} =$ $\sqrt{9} =$

$\sqrt{49} =$ $\sqrt{144} =$

$\sqrt{25} =$ $\sqrt{100} =$

$\sqrt{36} =$ $\sqrt{121} =$

D. Evaluate each expression. Review the order of operations in Lesson 138 if needed.

$\sqrt{36} \times \sqrt{4} - \sqrt{81} \div 3^2 = 6 \times 2 - 9 \div 9 = 12 - 1 = 11$

$(7 - \sqrt{16})^2 \times \sqrt{25} =$

$2^3 \div \sqrt{64} - \sqrt{9} + 9^1 =$

$\sqrt{10000} \div 5^2 + 4^2 \div 2^4 =$

Easy Peasy All-in-One Homeschool   EP Math 5/6 Workbook

LESSON 139 Practice

A. Make your own square root chart.

= 1	= 6	$\sqrt{121} = 11$	= 16
= 2	= 7	= 12	$\sqrt{289} = 17$
$\sqrt{9} = 3$	= 8	= 13	= 18
= 4	= 9	= 14	= 19
= 5	$\sqrt{100} = 10$	= 15	= 20

B. Evaluate each expression.

$\sqrt{25} \times \sqrt{9} - \sqrt{64} \div 4 =$

$\sqrt{81} \times (\sqrt{16} - 5)^2 =$

$\sqrt{64} \times \sqrt{100} + 1^5 =$

$2^5 \div 2^3 - \sqrt{36} + \sqrt{4} =$

$(\sqrt{49} - 6)^3 \div 3^1 \times 3^3 =$

C. What number am I? Use the clues to solve each problem.

I am a positive integer. My square is equal to my square root. What number am I? _____

I am a 2-digit square number. The sum of my digits is my square root. What number am I? _____

My square root is a prime number less than 15. All of my digits are prime numbers. What number am I? _____

I am a 2-digit number divisible by 2, 3, and 5. The sum of my digits is a square number. What number am I? _____

Easy Peasy All-in-One Homeschool   EP Math 5/6 Workbook

Made in United States
Orlando, FL
13 August 2022